Global Security and Intelligence Studies

Also from Westphalia Press
westphaliapress.org

GLOBAL SECURITY & INTELLIGENCE STUDIES

Volume 8, Number 1 • Summer 2023

Carter Matherly , Matthew Loux & Jim Burch, editors

Westphalia Press

An imprint of Policy Studies Organization

GLOBAL SECURITY AND INTELLIGENCE STUDIES
VOLUME 8, NUMBER 1 • SUMMER 2023

Westphalia Press
An imprint of Policy Studies Organization
1367 Connecticut Avenue NW
Washington, D.C. 20036
info@ipsonet.org

ISBN: 978-1-63723-727-4

Interior design by Jeffrey Barnes
jbarnesbook.design

Daniel Gutierrez-Sandoval, Executive Director
PSO and Westphalia Press

Updated material and comments on this edition
can be found at the Westphalia Press website:
www.westphaliapress.org

Making an Impact on Thought Leadership and Scholarship

The Journals of American Public University System

American Public University System (APUS) inspires scholars across the globe through its collection of peer-reviewed journals:

- Space Education and Strategic Applications
- Global Security and Intelligence Studies
- International Journal of Open Educational Resources
- Journal of Online Learning Research and Practice
- Saber and Scroll Historical Journal

A pioneer in online higher education, APUS—through American Military University and American Public University—ensures that working professionals are provided the opportunity to take full advantage of a quality affordable education.

Delve into our wide-ranging selection of journals and you'll understand why 200,000+ students and alumni have chosen APUS to help them achieve their scholastic objectives.

APUS—Defining Educational Foundations
Learn more at apus.edu/journals

Global Security and Intelligence Studies strives to be the source for research on global security and intelligence matters. As the global threat-scape evolves over time, *GSIS* is evolving to keep pace. The journal is enhancing its academic edge, impact, and reach. We are working to build stronger bridges between senior leaders, academics, and practitioners. In addition to new content that advances the global discussion of security and intelligence, readers can anticipate more special issues with a focus on current security concerns.

Global Security and Intelligence Studies is one of five journals sponsored by American Public University and published by Policy Studies Organization. The *International Journal of Online Educational Resources* (IJOER) publishes academic research with an emphasis on representing Open Educational Resources in teaching, learning, scholarship, and policy. The *Journal of Online Learning Research and Practice* (JOLRAP) publishes articles that focus on aspects related to virtual instruction, technology integration, data, ethics, privacy, leadership, and more. *Space Education and Strategic Applications* (SESA) journal encourages the publication of advances in space research, education, and applications. And lastly, *The Saber and Scroll* is a student- and alumni-led journal that publishes a variety of research on history and military history topics, book reviews, and exhibit/museum reviews.

Please visit https://www.apus.edu/academic-community/journals/index for more information on each journal.

A very special thank you to American Public University and the Policy Studies Organization for your generous and continued support of GSIS.

If you have research, notes, concepts, or ideas that you want to share, please do not hesitate to reach out with your submission! Our editorial team is always available to support new authors seeking to make an impact on the industry. Please visit us on Scholastica, https://gsis.scholasticahq.com/for-authors, for specifics on submissions.

Global Security and Intelligence Studies

Global Security and Intelligence Studies is published by The Policy Studies Organization on behalf of American Public University System. *GSIS* is licensed under a Creative Commons Attribution - NonCommercial - NoDerivatives 4.0 International License.

Aims and Scope. GSIS is a bi-annual, peer-reviewed, open-access publication designed to provide a forum for the academic community and the community of practitioners to engage in dialogue about contemporary global security and intelligence issues. The journal welcomes contributions on a broad range of intelligence and security issues, and from across the methodological and theoretical spectrum.

The journal especially encourages submissions that recognize the multidisciplinary nature of intelligence and security studies, and that draw on insights from a variety of fields to advance our understanding of important current intelligence and security issues. In keeping with the desire to help bridge the gap between academics and practitioners, the journal also invites articles about current intelligence and security-related matters from a practitioner perspective. In particular, *GSIS* is interested in publishing informed perspectives on current intelligence and security-related matters.

GSIS welcomes the submission of original empirical research, critical analysis, policy analysis, research notes, and book reviews. Papers and research notes that explicitly demonstrate how a multidisciplinary approach enhances theoretical and practical understanding of intelligence and security matters are especially welcome. Please visit: https://www. ipsonet.org/publications/open-access/gsis/instructions-for-authors

Global Security and Intelligence Studies
Volume 8, Number 1 • Summer 2023
© 2023 Policy Studies Organization

(cont'd.)

Notes From the Field

EDITORIAL WELCOME TO THE SPECIAL EDITION
Strategic Deterrence

Greetings! Welcome the start of our 8th volume and a significant turning point not only historically as this edition will explore, but for the *Global Security and Intelligence Studies* journal.

As you can tell, I am a tad excited and have some news to share, which I hope you will be just as excited about. Firstly, *Global Security and Intelligence Studies* was founded in 2015 to provide an outlet and collaboration point for scholars and practitioners to exchange ideas and experience. Over the past 8 years, the journal has grown in scope, readership, submissions, and credibility. Over the past year, we have done some "market research" on academic journals following trends, citability, discovery, and indexing. In order to maintain relevance and continue to grow the journal, you can expect to see a slight rebranding. In close consultation with APUS and PSO, we have taken steps to change the name of the journal to *Security and Intelligence* and will update our logo to reflect the streamlined feel. We believe that these small changes, combined with our growing readership and the highest level of scholarly work, will help the journal grow to new heights within the community. Thank you for being a part of the GSIS journey as we grow.

Over 75 years ago the first nuclear device was detonated, sparking a new era in strategic deterrence that would remain largely unchallenged for nearly a century. Biological warfare, chemical weapons, state sponsored terrorism, etc., have all shifted the discussion at different levels, but the measuring stick never departed from nuclear weaponry in a meaningful way. This stalwart standard effectively ensured a static status quo of hegemonic power during that time. However, hypersonic weapons, advanced cyber weaponry and effects, along with legal loopholes and financial shifting, have begun to shake that steady status quo. Examples of the beginning shift in power can be seen in Russia's invasion of Ukraine, China's aggressive rhetoric regarding Taiwan and the Philippines, Niger's coup, and the countless cyber-attacks targeting global interests across a wide variety of varying interests. This issue explores the facets of what strategic deterrence is and what the next evolution of deterrence will be.

doi: 10.18278/gsis.8.1.1

Graduate Lectern

Our Graduate Lectern this edition comes from William Hubbell with his piece "The Threat of China's MSS: American Universities, Corporations, and Overseas Intelligence Operations." In it, he details some of the activities and impacts attributed to China's Ministry of State Security's short-term and long-term exploitation of American universities, corporations, and overseas intelligence operations. His discussion of these "soft targets" highlights the multi domain and approach adversaries will take to ensure a contested environment exists, regardless of nationality or alliance.

Research Articles & Critical Analysis

Perhaps one of the most impactful modern advances in kinetic weaponry is the technological breakthrough in hypersonic weaponry. As new technology matures, paradigm shifts in how great powers deter and conduct warfare will mandate reevaluation of deterrence concepts and theory. They are extraordinarily powerful even without a warhead—the energy of a 20-kilogram brick at Mach 6 would have the same explosive force as a Mk-84 2,000lb bomb. Matt Dougherty explores the impact of Chinese operationalizing of such weapons in "The Dragon's Tail: Deterring China in an Era of Maneuverable Hypersonic Weapons." This research presents and analyzes three distinct options for U.S. deterrence theory against five Joint Planning Process evaluation criteria in a timely and critical time.

As mentioned in the introduction, cyber effects have contributed considerably to the calculus of strategic deterrence. Yang Liu dives deep into past and developing U.S. cyber security policy in "Doxfare as a Tool for Strategic Deterrence." Liu highlights the shortcomings of the 2018 policy of passive defense in favor of a modern effective tool for deterring hostile state cyber operations—doxfare. The work analyzes why doxfare is an effective deterrence strategy and addresses the ethical and legal concerns about doxfare.

Furthering the domain-oriented fight, our third article comes from Armin Krishnan and builds upon the emerging fifth-generation warfare concept in "Havana Syndrome: A Case of Fifth Generation Warfare." Throughout the history of warfare, the primary means of enacting combat has been through kinetic actions; the injuries sustained by U.S. consulate personnel stationed in Havana, Cuba, known as Havana Syndrome, carry the hallmarks of non-kinetic fifth generation warfare. Krishnan offers insights into the effects and intent of such attacks and how they work to undermine legitimacy of the state.

We then take a moment to rewind to the late 1930s, where Mona Parra gives and excellent recount and discussion on the importance of espionage activities in

"SNOW, Double Agent: The Beginnings of the Double Cross Network." It is critical to remember that despite all the technological and doctrinal advances of deterrence, human intelligence operations have always been one of the longstanding requirements in a deterrence strategy—acknowledged or not.

The root cause of most deterrence situations stems from some form of incongruities between nations. Historically, the Japanese and South Korean nations have been victim to these incongruities. Kenneth Sandler explores international relations theories as applied to these two nations in "Competitive, Cooperative, and Confounding Japan and South Korean relations through International Relations Theories." Realism posits that nations with common interests or enemies are likely to be partners. Part of strategic deterrence is maintaining strong allies and coalitions. Sandler's work sheds needed light on the complex problems that may arise from entangling alliances and histories.

Our last collection of articles in this issue highlight shifts seen in Russian military tactics. First Christina Roberts & Keith Ludwick, examine three high visibility and high impact cases—Cambridge Analytica, Russian interference in the 2016 U.S. Presidential Election, and the January 6th insurrection at the U.S. Capitol, during which individuals' personal data was used to spread misinformation and disinformation. Their work "Social Media Lies: User's Private Information and the Spread of Misinformation and Disinformation" highlights the dangers of the collection and subsequent misuse of personal data. Second Eugene Vertlieb (with Dennis Faleris) Their second contribution, "The Cognitive-Strategic Meaning of the Special Military Operation in Ukraine," explores the strategic value of the "limited military operation" in Ukraine to the survival of the modern Russian ideology. Lastly Dr. Robert Girod presents an enlightening discussion of the strategic nature of border security with significant tiebacks to the invasion of Ukraine in "Border Security: Vulnerability And Emerging Global Security Threats".

Notes From the Field

The events in Ukraine and Russia continually evolve as the front lines and tactics shift seemingly by the hour. Eugene Vertlieb (with Dennis Faleris) brings us another phenomenal discussion. In "The Essence of the Prigozhin Phenomenon," we are presented with very detailed insight into the recent attempted coup. Collectively, their work is an essential read for any individual following the ongoing conflict.

Our last offering in our special edition comes from Repez Romeo-Ionuț and his exploration of a longstanding intelligence concept "Black Swan in Intelligence." While a "tried and true" concept in the industry, Romeo-Ionuț offers a different look at the use and impact of the concept that is worth reading and considering for all organizations in the industry.

It has been our pleasure and honor to deliver a cornerstone of academic thought on Strategic Deterrence. We are looking forward to our coming winter issue, in which Dr. Slonopas and Dr. Drumhiller present an insightful look into Cybersecurity and Strategic Implications. Thank you to all our supporters, readers, authors, the American Public University System, and the Policy Studies Organization. A very special thank you to our reviewers and editorial leads for helping to make our journal what it is today.

Carter Matherly, Ph.D.
Co-Editor in Chief

Matthew Loux, Ph.D.
Co-Editor in Chief

Jim Burch, D.M.
Associate Editor

Editorial de bienvenida a la edición especial: Disuasión estratégica

¡Saludos! Le damos la bienvenida al comienzo de nuestro octavo volumen y un importante punto de inflexión no sólo históricamente, como explorará esta edición, sino también para la revista *Global Security and Intelligence Studies*.

Como podrá ver ver, estoy un poco emocionado y tengo algunas noticias que compartir, que espero que a usted también le entusiasmen. En primer lugar, *Global Security and Intelligence Studies* se fundó en 2015 para proporcionar un punto de salida y colaboración para que académicos y profesionales intercambien ideas y experiencias. Durante los últimos 8 años, la revista ha crecido en alcance, número de lectores, presentaciones y credibilidad. Durante el año pasado, realizamos algunas "investigaciones de mercado" sobre revistas académicas siguiendo tendencias, citabilidad, descubrimiento e indexación. Para mantener la relevancia y seguir haciendo crecer la revista, puede esperar ver un ligero cambio de marca. En estrecha consulta con APUS y PSO, hemos tomado medidas para cambiar el nombre de la revista a Seguridad e Inteligencia y actualizaremos nuestro logotipo para reflejar la sensación simplificada. Creemos que estos pequeños cambios, combinados con nuestro creciente número de lectores y el más alto nivel de trabajo académico, ayudarán a que la revista alcance nuevas alturas dentro de la comunidad. Gracias por ser parte del viaje de GSIS a medida que crecemos.

Hace más de 75 años se detonó el primer dispositivo nuclear, lo que inició una nueva era en la disuasión estratégica que permanecería prácticamente indiscutida durante casi un siglo. La guerra biológica, las armas químicas, el terrorismo patrocinado por el Estado, etc., han cambiado la discusión en diferentes niveles, pero la vara de medir nunca se apartó del armamento nuclear de manera significativa. Este estándar incondicional aseguró efectivamente un status quo estático de poder hegemónico durante ese tiempo. Sin embargo, las armas hipersónicas, el armamento y los efectos cibernéticos avanzados, junto con las lagunas legales y los cambios financieros, han comenzado a sacudir ese estable status quo. Se pueden ver ejemplos del cambio inicial en el poder en la invasión rusa de Ucrania, la retórica agresiva de China con respecto a Taiwán y Filipinas, el golpe de estado de Níger y los innumerables ataques cibernéticos dirigidos a intereses globales de una amplia variedad de intereses. Este número explora las facetas de lo que es la disuasión estratégica y cuál será la próxima evolución de la disuasión.

Atril de posgrado

Nuestro atril de posgrado de esta edición proviene de William Hubbell con su artículo "La amenaza de los MSS de China: universidades, corporaciones y operaciones de inteligencia en el extranjero estadounidenses". En él, detalla algunas de las actividades y los impactos atribuidos a la explotación a corto y largo plazo por parte del Ministerio de Seguridad del Estado de China de las universidades, corporaciones y operaciones de inteligencia estadounidenses en el extranjero. Su análisis de estos "objetivos fáciles" destaca los múltiples dominios y el enfoque que adoptarán los adversarios para garantizar que exista un entorno en disputa, independientemente de su nacionalidad o alianza.

Artículos de investigación y análisis crítico

Quizás no de los avances modernos más impactantes en armamento cinético sea el avance tecnológico en armamento hipersónico. A medida que madure la nueva tecnología, los cambios de paradigma sobre cómo las grandes potencias disuaden y conducen la guerra exigirán una reevaluación de los conceptos y la teoría de la disuasión. Son extraordinariamente poderosos incluso sin una ojiva: la energía de un ladrillo de 20 kilogramos a Mach 6 tendría la misma fuerza explosiva que una bomba Mk-84 de 2.000 libras. Matt Dougherty explora el impacto de la puesta en funcionamiento de tales armas por parte de China en "La cola del dragón: disuadiendo a China en una era de armas hipersónicas maniobrables". Esta investigación presenta y analiza tres opciones distintas para la teoría de la disuasión estadounidense frente a cinco criterios de evaluación del Proceso de Planificación Conjunta en un momento oportuno y crítico.

Como se mencionó en la introducción, los efectos cibernéticos han contribuido considerablemente al cálculo de la disuasión estratégica. Yang Liu profundiza en el pasado y en el desarrollo de la política de seguridad cibernética de Estados Unidos en "Doxfare como herramienta para la disuasión estratégica". Liu destaca las deficiencias de la política de defensa pasiva de 2018 en favor de una herramienta moderna y eficaz para disuadir las operaciones cibernéticas estatales hostiles: el doxfare. El trabajo analiza por qué la doxfare es una estrategia de disuasión eficaz y aborda las preocupaciones éticas y legales sobre la doxfare.

Nuestro tercer artículo, que promueve la lucha orientada a los dominios, proviene de Armin Krishnan y se basa en el concepto emergente de guerra de quinta generación en "El síndrome de La Habana: un caso de guerra de quinta generación". A lo largo de la historia de la guerra, el medio principal para realizar el combate ha sido mediante acciones cinéticas; Las lesiones sufridas por el personal del consulado estadounidense estacionado en La Habana, Cuba, conocidas como Síndrome de La Habana, llevan las características de la guerra no cinética de quinta generación. Krishnan ofrece información sobre los efectos y la intención de tales ataques y cómo funcionan para socavar la legitimidad del Estado.

Luego nos tomamos un momento para retroceder hasta finales de la década de 1930, donde Mona Parra ofrece un excelente recuento y discusión sobre la importancia de las actividades de espionaje en "SNOW, Double Agent: The Beginnings of the Double Cross Network". Es fundamental recordar que a pesar de todos los avances tecnológicos y doctrinales de la disuasión, las operaciones de inteligencia humana siempre han sido uno de los requisitos de larga data en una estrategia de disuasión, reconocida o no.

La causa fundamental de la mayoría de las situaciones de disuasión surge de algún tipo de incongruencia entre naciones. Históricamente, las naciones japonesa y surcoreana han sido víctimas de estas incongruencias. Kenneth Sandler explora las teorías de las relaciones internacionales aplicadas a estas dos naciones en "Las relaciones competitivas, cooperativas y confusas entre Japón y Corea del Sur a través de teorías de las relaciones internacionales". El realismo postula que es probable que naciones con intereses comunes o enemigos sean socios. Parte de la disuasión estratégica es mantener aliados y coaliciones fuertes. El trabajo de Sandler arroja luz sobre los complejos problemas que pueden surgir al entrelazar alianzas e historias.

Nuestra última colección de artículos en este número destaca los cambios observados en las tácticas militares rusas. Primero, Christina Roberts y Keith Ludwick examinan tres casos de gran visibilidad y alto impacto: Cambridge Analytica, la interferencia rusa en las elecciones presidenciales de EE. UU. de 2016 y la insurrección del 6 de enero en el Capitolio de EE. UU., durante la cual se utilizaron datos personales de personas para difundir información errónea y desinformación. Su trabajo "Mentiras en las redes sociales: información privada del usuario y la difusión de información errónea y desinformación" destaca los peligros de la recopilación y el posterior uso indebido de datos personales. Segundo Eugene Vertlieb (con Dennis Faleris) Su segunda contribución, "El significado cognitivo-estratégico de la operación militar especial en Ucrania", explora el valor estratégico de la "operación militar limitada" en Ucrania para la supervivencia de la ideología rusa moderna. Por último, el Dr. Robert Girod presenta una discusión esclarecedora sobre la naturaleza estratégica de la seguridad fronteriza con importantes vínculos con la invasión de Ucrania en "Seguridad fronteriza: vulnerabilidad y amenazas emergentes a la seguridad global".

Notas del campo

Los acontecimientos en Ucrania y Rusia evolucionan continuamente a medida que las líneas del frente y las tácticas cambian aparentemente cada hora. Eugene Vertlieb (con Dennis Faleris) nos trae otra discusión fenomenal: "La esencia del fenómeno Prigozhin", se nos presenta una visión muy detallada del reciente intento de golpe. En conjunto, su trabajo es una lectura esencial para cualquier persona que siga el conflicto en curso.

La última oferta de nuestra edición especial proviene de Repez Romeo-Ionuț y su exploración de un concepto de inteligencia de larga data, el "Cisne negro en la inteligencia". Si bien es un concepto "probado y verdadero" en la industria, Romeo-Ionuț ofrece una mirada diferente al uso y el impacto del concepto que vale la pena leer y considerar para todas las organizaciones de la industria.

Ha sido un placer y un honor para nosotros presentar una piedra angular del pensamiento académico sobre la disuasión estratégica. Esperamos con ansias nuestra próxima edición de invierno, en la que el Dr. Slonopas y el Dr. Drumhiller presentan una mirada profunda a la ciberseguridad y sus implicaciones estratégicas. Gracias a todos nuestros seguidores, lectores, autores, al Sistema Universitario Público Estadounidense y a la Organización de Estudios Políticos. Un agradecimiento muy especial a nuestros revisores y líderes editoriales por ayudar a hacer de nuestra revista lo que es hoy.

编者按
战略威慑

欢迎阅读《全球安全与情报研究》（GSIS）杂志第八卷，本期将探讨的主题是历史转折点，同样，本期也是本刊的转折点。

如您所知，我怀着些许兴奋来分享一些消息，并希望您也对此感到兴奋。第一，《全球安全与情报研究》成立于2015年，旨在为学者和从业人员提供交流思想和经验的平台和协作点。过去8年，本刊的研究范围、读者群、投稿量和可信度都在不断增长。 去年，我们对学术期刊追踪趋势、可引用性、发现和索引进行了一些"市场研究"。为了保持相关性并继续发展本刊，您将预期看到本刊更名。与美国公立大学系统(APUS)和政策研究组织(PSO)密切协商后，我们已采取步骤将期刊名称更改为《安全与情报》，并将更新我们的徽标以体现精简之感。我们相信，这些微小的变化，加上我们不断增长的读者群和最高水平的学术工作，将帮助本刊在研究界内发展到新的高度。感谢您参与GSIS的成长之旅。

75多年前，第一个核装置被引爆，开启了战略威慑的新时代，这一时代在近一个世纪内基本未受到挑战。生物战、化学武器、国家支持的恐怖主义等都在不同层面上改变了讨论，但衡量标准却从未以有意义的方式偏离核武器。这一坚定的标准有效保证了当时霸权的静态现状。然而，高超音速武器、先进的网络武器和效果，以及法律漏洞和金融转移，这一切已经开始动摇稳定的现状。权力开始转移的例子包括俄罗斯入侵乌克兰、中国对台湾和菲律宾的侵略性言论、尼日尔政变以及针对全球利益的无数网络攻击。本期探究了什么是战略威慑以及威慑的下一步演变是什么。

研究生讲台

本期的"研究生讲台"版块来自William Hubbell的文章《中国国家安全部的威胁：美国大学、企业和海外情报操作》。他在文章中详细介绍了中国国家安全部对美国大学、企业和海外情报操作进行的一系列短期及长期利用所产生的部分活动及影响。他对这些"软目标"的讨论强调了对手将采取多领域方法来确保竞争环境的存在，这与国籍或联盟无关。

研究文章与批判性分析

也许动能武器领域中最具影响力的现代进步之一是高超音速武器的技术突破。随着新技术的成熟，大国在阻止和发动战争的方式上所出现的范式转变将要求重新评价威慑概念及理论。即使没有弹头，这些武器的威力也非常强大——20公斤重的砖块在6马赫时的能量将与Mk-84 2,000磅炸弹的爆炸力相同。Matt Dougherty在《惹麻烦：在可操纵高超音速武器时代下威慑中国》一文中探究了中国对此类武器的操作化所产生的影响。这项研究在及时和关键的时刻根据5项联合规划过程(Joint Planning Process)评价标准，提出并分析了美国威慑理论的三种独特选择。

正如导论中提到的，网络效应对战略威慑的计算作出了巨大贡献。Yang Liu在《Doxfare作为战略威慑工具》一文中研究了美国网络安全政策的过去与现在。Liu强调了2018年被动防御政策的缺点，并支持采用doxfare这一现代有效工具来阻止敌对国家的网络行动。文章分析了为何doxfare是一种有效的威慑策略，并研究了有关doxfare的伦理及法律关切。

我们收录的第三篇文章由Armin Krishnan撰写，进一步研究了领域导向的战斗。文章题为《哈瓦那综合征：第五代战争案例》，并以新兴的第五代战争概念为基础。纵观战争史，发动战斗的主要手段一直是通过动能行动；驻古巴哈瓦那的美国领事馆人员所受的伤害被称为"哈瓦那综合征"，其带有非动能第五代战争的特征。Krishnan深入探讨了此类攻击的影响和意图，以及它们如何破坏国家的合法性。

下一篇文章回到20世纪30年代末，文章题为《双重特工SNOW：背叛网络的开端》，由Mona Parra撰写。作者对间谍活动的重要性进行了精彩的叙述和讨论。需要牢记的是，尽管威慑取得了技术进步和理论进步，但人类情报行动始终是威慑战略中长期存在的要求之一——无论承认与否。

大多数威慑局势的根本原因源于国家之间某种形式的不一致。从历史上看，日本和韩国一直是这些不一致的受害者。Kenneth Sandler在《透过国际关系理论看待日韩的竞争、合作与困惑关系》一文中探究了适用于这两个国家的国际关系理论。现实主义认为，具有共同利益或敌人的国家很可能成为合作伙伴。战略威慑的一部分是维持强大的盟友和联盟。Sandler的文章阐明了因纠缠的联盟和历史而可能产生的复杂问题。

本期收录的最后几篇文章聚焦于俄罗斯军事战术的转变。Christina Roberts和Keith Ludwick在其文章中研究了三个具有高知名度和高影响力的案例："剑桥分析"事件、俄罗斯干预2016年美国总统选举、以及1·6美国国会暴乱事件，在此期间，个人的私人数据被用来传播错误信息和虚假信息。他们的文章《社媒谎言：用户私人信息与错误及虚假信息的传播》强调了收集和滥用个人数据一事的危险。Eugene Vertlieb（与Dennis Faleris）撰写的第二篇文章题为《乌克兰特别军事行动的认知战略意义》，探究了乌克兰"有限军事行动"对现代俄罗斯意识形态生存的战略价值。最后，Robert Girod博士在《边境安全：

脆弱性与新兴的全球安全威胁》一文中对边境安全的战略性质进行了富有启发性的讨论，后者与入侵乌克兰一事有关。

领域观点

前线和战术似乎随时都在变化，据此，乌克兰和俄罗斯的冲突事件不断演变。Eugene Vertlieb（与Dennis Faleris）为我们带来了另一篇精彩的讨论文——《普里戈任现象的本质》，这篇文章为我们带来了有关近期未遂政变的详细见解。总的来说，他们的文章对于关注该持续冲突的所有人来说都是必读之物。

本期收录的最后一篇文章由Repez Romeo-Ionu撰写，他探究了长期存在的情报概念——"情报中的黑天鹅"。虽然这是一个"久经考验且真实"的概念，但Romeo-Ionus从不同的视角分析了该概念的使用和影响，这值得业内组织加以阅读和考量。

我们很高兴也很荣幸就战略威慑提供学术思想基石。我们期待今年的冬季期刊，届时Slonopas博士和Drumhiller博士将深入研究网络安全与战略启示。我们感谢本刊的所有支持者、读者、作者、美国公立大学系统和政策研究组织。特别感谢我们的审稿人和编辑领导——你们帮助本刊取得了今天的成就。

The Threat of China's MSS: American Universities, Corporations, and Overseas Intelligence Operations

William Hubbell

william.hubbell@mycampus.apus.edu

The intelligence community in the United States is widely regarded as one of the most advanced in the world. The Central Intelligence Agency (CIA), the National Security Agency (NSA), and others have long played a vital role in protecting the nation's national security interests, especially amid growing totalitarianism and extremist worldwide. Unfortunately, several countries wish harm upon the United States and its allies, which means that their respective intelligence communities tend to be fixated upon objectives detrimental to American national security interests. In light of the current war in Ukraine, alongside longstanding tension from the Cold War, the Komitet Gosudarstvennoy Bezopasnosti (KGB), otherwise known as the principal intelligence agency of Russia, has understandably garnered substantive interest from scholars and analysts alike. However, the Ministry of State Security (MSS) in China should be of significant concern to the United States, especially when accounting for the insidious ways in which the MSS can harm vital national security interests. Dorfman (2020) notes that the ongoing battle regarding data, namely "who controls it, who secures it, who can steal it, and how it can be used for economic and security objectives," has gradually come to define the growing conflict between Washington and Beijing, which in turn underscores the growing threat MSS poses to the United States. As detailed in the following analysis, China's MSS poses a significant threat to United States national security in the short-term and long-term due to its exploitation of American universities, corporations, and overseas intelligence operations.

The American university system constitutes an increasingly weak link in national security, especially considering how deeply integrated many of its services and programs are with Chinese interests. In general, universities across the United States "have long played a leading role in relations between the United States and China" (Diamond & Schell, 2019), especially in terms of the numerous students and professors of Chinese origin present across universities today. In light of several highly publicized incidents regarding the theft of technology or other serious issues, fears have risen regarding MSS's propensity for "using American universities as vehicles through which to advance Chinese Community Party propaganda" (Diamond & Schell, 2019). It is import-

doi: 10.18278/gsis.8.1.2

ant to note that Chinese law is remarkably strict in terms of the information that it is allowed to collect from its citizens, which means that it has access to virtually all student data and information while they study in the United States. In several cases, students may willingly provide data to MSS, especially if they enrolled in American universities for espionage purposes in the first place.

For this reason, American university system has become an optimal "soft target" in terms of the "the global espionage war with China" (Dilinian, 2020). As observed by Bill Evanina, who serves as one of the top counterintelligence officials in the United States: "A lot of our ideas, technology, research, innovation is incubated on those university campuses … That's where the science and technology originates – and that's why it's the most prime place to steal" (Dilinian, 2020). Universities, in other words, are a valuable source of data for MSS, especially as it continues contributing information to the growing, vast apparatus of AI driven technologies. In essence, "data has already critically shaped the course of Chinese politics," which in return has started "altering the course of U.S. foreign policy and intelligence gathering around the globe" (Dorfman, 2020). In remarks to *Foreign Policy*, Evanina illustrates the grave threat China poses in that it is "one of the leading collectors of bulk personal data around the globe, using both illegal and legal means" (Dorfman, 2020), which has been evident in several different cases.

These cases entail both students and professors who have been accused of providing information to or otherwise taking actions to benefit MSS. For instance, a Boston University student failed to disclose her position as a lieutenant in the People's Liberation Army (Dilanian, 2020). In addition, at the Illinois Institute of Technology in Chicago, a Chinese student was charged for attempting "to recruit spies for his country's version of the CIA" (Dilanian, 2020). Bo Mao, who was a professor at the University of Texas, stole proprietary technology from an American Silicon Valley startup and subsequently passed it along to Huawei, the highly controversial Chinese telecommunications giant. Mao's affiliation with the university is precisely what enabled him to steal the technology: "By using his status as a university researcher to obtain the circuit board under the guise of academic testing" (Dilanian, 2020). In other words, academics have the ability to gain access to otherwise privileged or confidential information under the name of research. This information can be highly secretive for a reason, especially if it pertains to nuclear technology or other highly specialized information. Moreover, academics can also participate directly in helping Chinese bolster its intelligence, further weakening national security.

One of the most egregious cases included that of Dr. Charles Lieber, the former Chair of Harvard University's Chemistry and Chemical Biology Department, who was charged alongside two Chinese nationals for aiding the People's Republic of China. Lieber's arrest occurred after it emerged that he had lied about his involvement with the Thousand Talents Plan, which has drawn increased scrutiny from the American intelligence community (Department of Justice, 2020). According to the Department of Justice (2020), China's Thousand Talents Plan constitutes "one of the most prominent Chinese Talent recruit plans that are designed to attract, recruit, and cultivate high-level scientific talent," chiefly to further "China's scientific development, economic prosperity and national security." In essence, this talent program "[seeks] to lure Chinese overseas talent and foreign experts to bring their knowledge and experience to China," as well as "reward individuals for stealing proprietary information" and providing it to Chinese intelligence authorities (Department of Justice, 2020). The program has drawn the great ire of American agencies, who have viewed it as a vehicle for the MSS to obtain information for purposes of furthering China's power. Lieber was involved in the Wuhan University of Technology, which paid him exorbitant compensation for his services: "WUT paid Lieber $50,000 USD per month, living expenses of up to 1,000,000 Chinese Yuan (approximately $158,000 USD at the time) and awarded him more than $1.5 million to establish a research lab at WUT" (Department of Justice, 2020). The level of greed exhibited by Lieber illustrates precisely how the MSS is able to exploit weak links in the American university system, and it continues this exploitative practice with regards to American corporations and their innovative technologies.

Whereas two decades ago a major concern was the targeted attack on classified Department of Defense websites, the major concern now includes "the shift [in espionage efforts] to private sector intellectual property research and development, particularly by China, who has been the most egregious one in stealing those technologies" (CBS News, 2021). Consequently, corporations, like universities, have become of great interest to MSS, especially since "another way to get the tiger is to circumvent the developmental process altogether by stealing the product" (Hannas et al., 2013). The espionage capabilities exhibited by MSS illustrate precisely why the NSA finds cyber espionage to be particularly damaging in terms of monetary loss. On July 26, 2012, General Keith Alexander of the NSA informed the Aspen Security Forum that cyber espionage constituted "the greatest transfer of wealth in history" (Hannas et al., 2013). Per the American government, U.S. corporations routinely "lose billions of dollars' worth of technological innovation each year to China" (Schnell, 2022), a staggering figure not only from the massive amount alone, but also the implications of that much proprietary technology being lost to China through its nefarious spy agencies' efforts. In other words, the real losses may be even greater if the United States loses its edge in innovation relative to China.

Multiple cases of Chinese executives or professionals engaging in crime on behalf of Chinese intelligence have abounded, especially when incentivized. For example, Xiangdong Yu had previously worked as a product engineer for Ford Motor Company, where he copied approximately 4,000 Ford documents into an external hard drive for purposes of obtaining a job with an automotive company in China (Hannas et al., 2013). However, Yu was captured in October 2009 and pled guilty to one count of theft of trade secrets (Hannas et al., 2013). In one especially remarkable case, an actual MSS officer managed to penetrate an American corporation and attempt to obtain highly sensitive information. Yanjun Xu, under the direction of the MSS, was "accused of seeking to steal General Electric/Aviation jet engine technology" (Schnell, 2022), which, given the potential military applications, constitutes a serious issue. Moreover, Xu's case was highly unique in the sense that it employed remarkable cyber subterfuge efforts: "A unique feature of his case is that it he allegedly did so without ever setting foot in the [United States]" (Schnell, 2022). In other words, as Chinese capabilities in cyber espionage continue to advance, the technology of American companies is increasingly threatened. This threat is compounded by the reality that Americans and Chinese are deeply integrated in technological innovation and production, which is why Former Secretary of Defense Robert Gates advocated a "small yard, high fence" approach for protecting American corporations. Specifically, Gates called for "selectively protecting key technologies, and doing so aggressively" (Hass & Balin, 2019) in an effort to protect American security interests.

While MSS can cause havoc internally in the United States, it can also create havoc for U.S. intelligence agencies externally, or over the course of their overseas intelligence operations. This level of disruption, along with domestic disturbances, is precisely why "the FBI opens a new China-related counterintelligence investigation every 12 [hours]" (Schnell, 2022) As of 2020, at least 5,000 cases were active (Wray, 2020). These cases also account for the fact that the MSS and its support have access to highly sophisticated technologies that cause a serious threat to United States operations overseas. For example, in 2013, American intelligence agencies discovered a highly troubling trend: "Undercover CIA personnel, flying into countries in Africa and Europe for sensitive work, were being rapidly and successfully identified by Chinese intelligence," and "in some cases as soon as the CIA officers had cleared passport control" (Dorfman, 2020). In general, American intelligence attempts to recruit "Russians and Chinese hard in Africa" (Dorfman, 2020) per one former official, which is precisely why the exposure of these agents is highly problematic in terms of their safety.

Moreover, when recalling the troves of data MSS and other agencies have managed to obtain via hacks, it is important to note that, "compounding the threat, the data China stole is of obvious value as they attempt to identify people for secret intelligence gathering" (Wray, 2020). Specifically, with regards to MSS's ability to identify CIA agents as soon as they cleared passport control, "U.S. officials believed Chinese intelligence operatives had likely combed through and synthesized information from these massive, stolen caches to identify the undercover U.S. intelligence officials" (Dorfman, 2020). A former intelligence official referred to the "suave and professional utilization" of these data sets as neither "random," nor "generic," but rather "a big data problem" (Dorfman, 2020). This situation illustrates precisely why "rapid escalation [constitutes] an acute risk, particularly if the pace of technological advancements in capabilities exceeds the development of protocols for maintaining human agency in decision-making loops" (Hass & Balin, 2019). In other words, whereas intelligence agencies could try to root out moles in the past, big data analytics, which may be more effective than moles, is a far more difficult foe to defeat.

In general, China arguably poses the biggest threat, or clearly one of the biggest threats, to the United States in a number of ways: Per Joseph Bonavolonta, an agent with the FBI, "no country poses a greater, more severe or long-term threat to our national security and economic prosperity than China" (Dilianian, 2020). Thus, its premiere intelligence agency, MSS, is a huge threat due to its ability to exploit American universities, corporations, and overseas intelligence operations. Theft of proprietary information from universities and corporations is commonplace, among other behaviors, some of which are directly executed by MSS officers themselves. Moreover, advancements in artificial intelligence are enabling the MSS to become an even fiercer opponent in overseas intelligence operations. As noted by Bonavolonta, "China's communist

government's goal, simply put, is to replace the U.S. as the world superpower, and they are breaking the law to get there" (Dilianian, 2020). However, it is important to note that "China's appetite for foreign technology and its network for informal technology acquisition extend well beyond the United States," including attacks on the UK and other allies as well (Hannas et al., 2013). Consequently, it is crucial for interagency collaboration to attempt countering the growing China threat posed through its intelligence agency, MSS.

To strengthen cybersecurity measures, there are several methods to counter China from posing threats in America. These include enhancing cybersecurity protocols within universities, corporations, and government agencies to safeguard against cyber threats and intellectual property theft. International cooperation to prevent China's MSS from taking over American universities requires a multi-faceted approach. Countries need to establish effective information sharing mechanisms, exchanging intelligence on MSS activities and individuals or organizations with suspected ties to the MSS. This collaborative effort can aid in identifying potential threats and devising countermeasures. Additionally, policy coordination is vital to safeguard the independence and integrity of academic institutions. By developing unified guidelines and regulations, countries can address vulnerabilities, increase transparency in research funding, and protect academic freedom, reducing the MSS's ability to exploit loopholes in the system.

Bolstering cybersecurity measures is essential in preventing MSS-related cyber espionage and intellectual property theft. International cooperation should focus on sharing best practices, conducting joint cybersecurity drills, and coordinating responses to cyber threats originating from MSS-affiliated entities. By leveraging collective expertise and resources, universities can enhance their cybersecurity infrastructure, minimizing the risk of MSS interference. Furthermore, academic exchanges and collaborations should continue, but with increased transparency and scrutiny. Establishing guidelines for vetting partnerships and research collaborations, particularly in sensitive areas, can help mitigate the risk of undue influence. Overall, international cooperation plays a crucial role in fortifying American universities against MSS infiltration while promoting academic freedom and knowledge sharing (Jinghua, 2019).

Another method is by encouraging China to adhere to international norms and rules governing cybersecurity and intellectual property protection. This involves a two-fold approach. Primarily, diplomatic engagement is essential. Open and constructive dialogue should be established to foster mutual understanding and emphasize the benefits of compliance. Diplomatic efforts should stress the significance of cybersecurity and intellectual property protection for global economic growth and stability, emphasizing the advantages of a level playing field and fair competition. Through sustained diplomatic engagement, China can be encouraged to recognize the importance of adhering to these norms.

Furthermore, incentives and cooperation are key. Positive incentives can motivate China to align its practices with international standards. These incentives may include trade and economic benefits tied to compliance, such as enhanced market access and preferential treatment. Bilateral and multilateral cooperation can facilitate collaboration on cybersecurity and intellectual property protection. Sharing best practices, expertise, and technology, as well as jointly addressing common challenges, can demonstrate the advantages of adhering to international norms. By fostering public-private partnerships and providing technical assistance, capacity-building initiatives can be undertaken to assist China in effectively addressing these issues. These combined efforts, encompassing incentives, cooperation, and capacity-building, can contribute to the gradual adoption of international norms and rules by China.

The 2020 Annual Report to Congress highlights the importance of industry-government partnerships in fostering closer collaboration between the government and private sector entities, particularly universities and corporations, to exchange information, insights, and technological expertise. These partnerships can facilitate the sharing of knowledge and resources, enabling the government to leverage the expertise and innovation of the private sector. By establishing robust channels for communication and cooperation, such partnerships can enhance the exchange of critical information on emerging technologies, cybersecurity threats, and best practices. The collaboration between universities and corporations can facilitate joint research initiatives, technology transfer, and workforce development programs, fostering innovation and driving economic growth. Ultimately, these industry-government partnerships can create a dynamic ecosystem that harnesses the strengths of both sectors, promoting information sharing, innovation, and technological advancement for the benefit of society as a whole. (Congress, 2020).

Lastly, engaging in international discussions and negotiations is vital to establish norms and agreements governing cyber operations, intelligence activities, and the protection of intellectual property. By participating in these discussions, countries can collectively address the challenges posed by China's non-traditional espionage activities. International discussions provide a platform for countries to exchange perspectives, share experiences, and develop common understandings of the threats posed by cyber operations and intellectual property theft. These discussions aim to establish international norms and agreements that outline acceptable behavior in cyberspace, define

the boundaries of intelligence activities, and promote the protection of intellectual property rights. Through negotiations, countries can work towards consensus on these issues, seeking to establish legally binding agreements or frameworks. These agreements can serve as a guide for responsible behavior in cyberspace, discourage malicious activities, and outline consequences for non-compliance. Additionally, discussions and negotiations can help shape the development of international standards and guidelines to address emerging threats and challenges. (Department of Justice, 2020).

It is important to note that these counter strategies should be tailored to the specific challenges posed by China's MSS and regularly evaluated and adapted to address evolving threats in the intelligence and cybersecurity landscape. Also, a holistic approach that combines technological advancements, intelligence sharing, public-private partnerships, and international cooperation to effectively counter the MSS threat. Constant vigilance, proactive measures, and adaptability are also key to mitigating risks and safeguarding national security interests in the face of evolving intelligence challenges from China.

References

Diamond, L. & Schell, O. (2019). *China's Influence and American Interests: Promoting Constructive Vigilance*. Hoover Institution Press. Dilanian, K. (2020). American universities are a soft target for China's spies, say U.S. intelligence officials. *NBC News*. https://www.nbcnews.com/news/china/american-universities-are-soft-target-china-s-spies-say-u-n1104291

Dorfman, Z. (2020). China Used Stolen Data to Expose CIA Operatives in Africa and Europe. *Foreign Policy*. https://foreignpolicy.com/2020/12/21/china-stolen-us-data-exposed-cia-operatives-spy-networks/

Jinghua, L. (2019, March 22). *What are China's cyber capabilities and intentions?* IPI Global Observatory. https://theglobalobservatory.org/2019/03/what-are-chinas-cyber-capabilities-intentions/

Hannas, W.C., Mulvenon, J., Puglisi, A.B. (2013). *Chinese Industrial Espionage: Technology Acquisition and Military Modernization*. Taylor & Francis.

Harvard University Professor and Two Chinese Nationals Charged in Three Separate China Related Cases. (2020). *Department of Justice*. https://www.justice.gov/opa/pr/harvard-university-professor-and-two-chinese-nationals-charged-three-separate-china-related

Hass, R. & Balin, Z. (2019). US-China relations in the age of artificial intelligence. *Brookings*. https://www.brookings.edu/research/us-china-relations-in-the-age-of-artificial-intelligence/

Schnell, J. (2022). Cultural Variables Within Prosecution of Chinese Corporate Espionage: The Case of USA Versus Yanjun Xu. *Fudan J. Hum. Soc. Sci.* https://doi.org/10.1007/s40647-022-00350-0

The Strategic Competition Act of 2021, 117th Cong., 1st Sess. (2021). https://www.foreign.senate.gov/imo/media/doc/DAV21598%20-%20Strategic%20Competition%20Act%20of%202021.pdf

Top counterintelligence official Mike Orlando on foreign espionage threats facing U.S. (2021). *CBS News*. https://www.cbsnews.com/news/foreign-espionage-threats-u-s-intelligence-matters-podcast/

U.S.-China Economic and Security Review Commission. (2020). *2020 annual report to Congress*. Retrieved May 26, 2023, from https://www.uscc.gov/sites/default/files/2020-12/2020_Annual_Report_to_Congress.pdf

U.S. Department of Justice. (2018, December 12). *China's non-traditional espio-nage against the United States: The threat and potential policy responses.* https://www.justice.gov/sites/default/files/testimonies/witnesses/attachments/2018/12/18/12-05-2018_john_c._demers_testimony_re_china_non-traditional_espionage_against_the_united_states_the_threat_and_potential_policy_responses.pdf

Wray, C. (2020). The Threat Posed by the Chinese Government and the Chinese Communist Party to the Economic and National Security of the United States. *FBI.* https://www.fbi.gov/news/speeches/the-threat-posed-by-the-chinese-government-and-the-chinese-communist-party-to-the-economic-and-national-security-of-the-united-states

The Dragon's Tail: Deterring China in an Era of Maneuverable Hypersonic Weapons

By Lt Col Matthew Dougherty, USAF

School of Advanced Nuclear Deterrence Studies & School of Advanced Air and Space Studies, Air University, Maxwell AFB, Alabama

matthew.dougherty.3@us.af.mil

DISCLAIMER

The article as it appears is a shortened version of the original thesis with the same title. The original thesis may be found on the Air University Library searchable database or by request.

The conclusions and opinions expressed in this thesis are those of the author and do not necessarily reflect the official policy or position of the U.S. Government, Department of Defense, or The Air University.

Abstract

Maneuverable hypersonic weapons can travel at five to twenty-five times the speed of sound and maneuver along pre-planned routes prior to striking their assigned target(s), unlike traditional ballistic missiles. These weapons are extraordinarily powerful even without a warhead, as the energy of a 20-kilogram brick at Mach 6 would have the same explosive force as a Mk-84 2,000lb bomb. As this new technology matures, paradigm shifts in how great powers deter and conduct warfare will mandate reevaluation of deterrence concepts and theory. This research presents and analyzes three distinct options for U.S. deterrence theory against five Joint Planning Process evaluation criteria. The study finds that developing and deploying a dual-capable Hypersonic Cruise Missile leverages mutual vulnerability and escalation dominance to mitigate the asymmetric benefit gained by China as the best course of action.

Keywords: Hypersonic Weapons, nuclear deterrence, great power competition, China, Warfare

doi: 10.18278/gsis.8.1.3

La cola del dragón: disuadir a China en una era de armas hipersónicas maniobrables

Resumen

Las armas hipersónicas maniobrables pueden viajar entre cinco y veinticinco veces la velocidad del sonido y maniobrar a lo largo de rutas planificadas previamente antes de alcanzar sus objetivos asignados, a diferencia de los misiles balísticos tradicionales. Estas armas son extraordinariamente poderosas incluso sin una ojiva, ya que la energía de un ladrillo de 20 kilogramos a Mach 6 tendría la misma fuerza explosiva que una bomba Mk-84 de 2.000 libras. A medida que esta nueva tecnología madure, los cambios de paradigma sobre cómo las grandes potencias disuaden y conducen la guerra exigirán una reevaluación de los conceptos y la teoría de la disuasión. Esta investigación presenta y analiza tres opciones distintas para la teoría de la disuasión estadounidense frente a cinco criterios de evaluación del Proceso de Planificación Conjunta. El estudio encuentra que el desarrollo y despliegue de un misil de crucero hipersónico de doble capacidad aprovecha la vulnerabilidad mutua y el dominio de la escalada para mitigar el beneficio asimétrico obtenido por China como el mejor curso de acción.

Palabras clave: Armas hipersónicas, disuasión nuclear, competencia entre grandes potencias, China, Guerra

惹麻烦：在可操纵高超音速武器时代下威慑中国

摘要

与传统弹道导弹不同，可操纵高超音速武器能以5到25倍的音速行进，并在打击指定目标之前沿着预先计划的路线进行操纵。即使没有弹头，这些武器的威力也非常强大，因为20公斤重的砖块在6马赫时的能量与Mk-84 2,000磅炸弹的爆炸力相同。随着这项新技术的成熟，大国在阻止和发动战争的方式上所出现的范式转变将要求重新评价威慑概念及理论。本研究根据5项联合规划过程(Joint Planning Process)评价标准，提出并分析了美国威慑理论的三种独特选择。研究发现，开发和部署双功能高超音速巡航导弹是最佳的行动方案，其能利用相互脆弱性和（冲突）升级优势来减轻中国获得的不对称利益。

关键词：高超音速武器，核威慑，大国竞争，中国，战争

Introduction

"The nature of war is immutable because it involves human beings. The character of war changes. Frequently, **oftentimes, by technologies,** sometimes by demographics or politics."

—General Mark Milley, 20th Chairman of the Joint Chiefs of Staff

During the recent commissioning of an Australian hypersonics research center, Australian Defense Minister Peter Dutton described the development of Maneuverable Hypersonic Weapons (MHWs) as "the most strategic change to the strategic environment since the end of World War II."[1] Further, the Missile Defense Agency (MDA) Director, Vice Adm John Hill, noted that though they can detect the initial launch of a maneuverable hypersonic weapon, the United States cannot currently track, target, or engage MHWs employed against the United States or its vital interests.[2] Such an exposed vulnerability is especially concerning as MHWs are "dual-capable" of either conventional or nuclear payloads and represent a new, exotic means of providing immediate regional or global strike that is entirely different from nuclear intercontinental ballistic missiles (ICBMs) and sub-sonic, air-breathing cruise missiles.

The central question is: *In an environment of long-term "Strategic Competition," under what conditions can the United States successfully deter China from employing dual-capable, maneuverable hypersonic weapons?* This question requires exploring classical and contemporary deterrence theory and considering the Chinese perspective regarding conventional and nuclear MHW development, purpose, and use of such weapons. Accordingly, this research offers deterrence posture options the United States should consider in forming a modern deterrence framework addressing the vulnerabilities presented by dual-capable MHWs employed against the U.S. interests. The purpose of this research is to investigate a deterrence solution until an available, cost-effective, and reliable solution is available to defend against Chinese employment of dual-capable MHWs against the United States or its interests.

By definition, MHWs are weapons capable of traveling at five to twenty-five times the speed of sound and maneuvering along pre-planned routing before striking their assigned target(s), unlike traditional ballistic missiles that follow a predictable ballistic profile from launch to impact[3] (Fig. 1). In this new class of weapon, there are three predominant types of delivery methods: air-breathing hypersonic cruise missile (HCM), boost-glide hypersonic glide vehicle (HGV), and fractional orbital bombardment system (FOBS).[4] First, an air-breathing HCM may be launched from an airplane, ship, or vehicle and utilizes a scramjet engine, which propels the cruise missile to hypersonic speeds. Historically, igniting such an engine is extremely difficult and described as lighting a match in a 2,000-mph wind storm.[5] Once the scramjet engine is "lit," the weapon can fly at low-level altitudes, using its speed and the curvature of the Earth to delay detection by early warning radars until just before reaching its target.[6] Second, the boost-glide HGV can be launched from an airplane, ship, or missile silo. To launch the weapon, the HGV payload is boosted by a rocket motor similar to a ballistic missile, and upon re-entry into Earth's atmosphere, the HGV payload is released, executes a "pull-up" maneuver, and aerodynami-cally skips unpowered along the upper atmosphere at hypersonic speeds and maneuvers in the "glide phase" towards its intended target before transitioning to the "terminal phase" (or delivery) of the strike.[7] Finally, the FOBS uses an ICBM-class rocket booster to place the HGV payload in a fractional low-earth orbit (i.e., less than one complete orbit), intentionally de-orbits, re-enters the atmosphere, and initiates the same "pull-up" maneuver accomplished by the non-orbital HGV weapon.

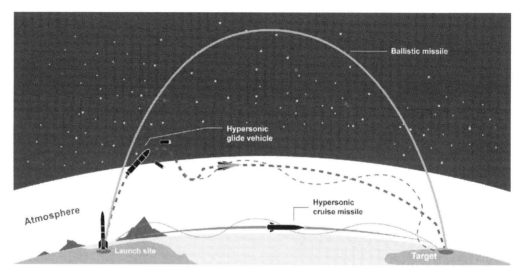

Figure 1. Maneuverable Hypersonic Weapon profiles compared to a traditional ballistic missile profile. Note: the FOBS profile is not depicted. (Reprinted from report, Shannon Bugos and Kingston Reif, Arms Control Association, subject: Understanding Hypersonic Weapons, September 2021.)

The Joint Chiefs of Staff Chairman, General Milley, recently described the Chinese FOBS test as a "Sputnik Moment."[8] The Chairman went further, describing China and Russia's recent MHW technological advances as unprecedented technological developments.[9] These comments echo Guilio Douhet's warning that the character of future conflict would be entirely different from previous conflict and that "victory smiles upon those who anticipate the changes in the character of war."[10] However, in light of the Chairman's beliefs and these prophetic words, the USAF has not incorporated MHWs into its Strategic Attack Doctrine.

This revolutionary technological advance presents the United States Government (USG) with the extreme challenge of defending against such weapons. Firstly, HCMs penetrate an A2/AD environment by exploiting a low-level approach with blistering speed and maneuverability, compressing an enemy's reaction time and ability to execute a meaningful defense. Secondly, an HGV employed by the FOBS is uniquely troubling because while space-based assets may easily detect its launch, it can exploit the north-facing, early-warning radar system by delivering the HGV on an avenue of approach "behind" the detectable region.[11] Moreover, in both cases of an HCM or HGV attack, the distinctive maneuverability of the weapon adds ambiguity to the intended target because traditional ballistic missile detection and tracking systems cannot predict where the weapon will hit. Adding to their complicated nature, MHWs can be extraordinarily powerful even without a warhead, as the energy of a 20-kilogram brick at Mach 6 would have the same explosive force as a Mk-84 2,000lb bomb.[12]

In response to the Chinese FOBS test, the Commander of NORAD and United States Northern Command, Gen Glen VanHerck, commented that Chinese capabilities would "provide significant challenges to my NORAD capability to provide threat warning and attack assessment."[13] Adm Charles Richard, commander of U.S. Strategic Command, recently testified to Congress that the FOBS weapon "flew over 25,000 miles for more than 100 minutes."[14] Douhet comically predicted attempts to shoot down an airplane, much like a ground-based interceptor might currently attempt to intercept

a MHW: "like a man trying to catch a homing pigeon by following him on a bicycle."[15] In a recent interview, Amb Robert A. Wood noted that "[w]e just do not know how we can defend against that [hypersonic] technology. Neither does China or Russia."[16] For these reasons, the problems created by these new MHWs as they relate to a new era of long-term "Strategic Competition" (formerly, "Great Power Competition").[17] Moreover, MHWs are currently uncontested by traditional ballistic missile defense (BMD) systems, given their unique maneuvering capability, blistering speed, and ability to exploit over-the-horizon early-warning blind spots. MHWs tend to operate "in the seams" between space-based sensors and ground-based radars, and even if the detection and tracking systems could effectively forewarn against such a weapon, the new U.S. Navy SM-6 GBI missile may not be able to engage an MHW in the terminal phase. Because of this, the MDA recently funded three GBI missile contracts and two Hypersonic and Ballistic Tracking Space Sensor (HBTSS) satellites to present a reliable defense by the decade's end.[18]

In his most recent publication on the return of great power rivalry, Dr. Matthew Kroenig echoed Douhet by saying, "[s]tates that push the technological frontier of military technology and, importantly, that can develop the new operational concepts to employ the technology on the battlefield, have an advantage over those that do not."[19] Unfortunately, in 2021, the former Vice Chairman of the Joint Chiefs of Staff, General John Hyten, acknowledged the United States is trailing China specifically by noting, "China has accomplished hundreds of hypersonic weapons tests compared to only 9 U.S. tests in the last five years."[20]

Table 1. Maneuverable Hypersonic Weapons Programs in Development

Weapon	Country	Type	Delivery Platform	Range (km)	Speed (Mach)	Nuc/Conv	IOC Date
DF-ZF (WU-14)	China	HGV	DF-17/DF-26	1,800-2,500	5-10M	Nuc/Conv	2020
FOBS	China	HGV	Long March-2F Y13	Unlimited	UNK	UNK	UNK
Xing Kong-2 (Starry Sky-2)	China	"Waverider"	UNK	UNK	6M	Nuc	2025
AVANGARD (Yu-71 or '4202')	Russia	HGV	ICBM	6,000+	20M+	Nuc	2019
TSIRKON (or ZIRCON)	Russia	HCM (anti-ship)	Submarine	500-1,000	5-8M	Conv	2023
KINZHAL (Kh-47M2)	Russia	ALBM	Airplane	2,000	10M	Nuc/Conv	2018
Hwasong-8	DPRK	MaRV (HGV?)	IRBM	700	10M	Nuc/Conv	2021
ARRW (Lockheed Martin)	US	HGV	Airplane	1,600	6.5-8M	Conv	UNK
HAWC/HACM (DARPA)	US	HCM	Airplane	UNK	5M+	Conv	UNK
USA LRHW (Dark Eagle)	US	HGV	Vehicle	2,775	5M+	Conv	2023
USN Conventional Prompt Strike	US	HGV	Destroyer/SSGN	UNK	5M+	Conv	2025/2028
Southern Cross Integrated Flight Research Experiment (SCIFiRE)	US/AUS	Boosted/A-B HCM	F-18/F-35/P-8	UNK	5M	Conv	2027-2032

Adapted from Shannon Bugos and Kingston Reif, Arms Control Association, subject: Understanding Hypersonic Weapons, September 2021.

As the world's front-runner on MHW developmental technology, China's current hypersonic weapons testing demonstrates unprecedented capability in hypersonic weapons development. Many have claimed (including the current Secretary of the Air Force, Mr. Frank Kendall) that the United States is now in a "hypersonic arms race" with China.[21] China is the top pacing challenge for the United States, and analysts predict China is intentionally adding MHWs to its deterrence posture to disrupt the U.S. strategic superiority mindset.[22] China showcased its premier hypersonic weapon, the DF-17, with a DF-ZF HGV payload during a parade in October 2019.[23] Mounted on a medium-range ballistic missile rocket similar to the DF-15, this new weapon system could serve as a dual-capable weapon to defeat U.S. missile defense systems. Of note, U.S. senior military leaders assess that only eight conventional DF-ZF HGV weapons would be needed to sink a U.S. aircraft carrier successfully.[24] China's FOBS test used an HGV deployed from a Long-March-2F Y13 rocket in their most recent test. This test was astounding because it likely released an unknown projectile from the HGV during the glide phase, which was previously considered an impossible task.[25] Finally, FOBS vehicles present an unexpected problem to U.S. early warning and missile defenses because they can approach from "behind" the early-warning, over-the-horizon radar system.

The United States re-ignited its pursuit of an air-breathing hypersonic cruise missile program from 2005-2010 in the form of the Boeing X-51 "Wave Rider." The "Wave Rider" served as a scramjet-powered hypersonic capability demonstrator but, unfortunately, was canceled after one successful test flight and a litany of previously failed flight tests. It was not until 2018 that the U.S. re-branded its pursuit of a hypersonic weapon through both the U.S. Air Force's AGM-183 Advanced Rapid Response Weapon (ARRW, pronounced "arrow"; boost-glide HGV) program and the Hypersonic Conventional Strike Weapon (HCSW, pronounced "hacksaw"; HCM). Both platforms are air-launched and have suffered under receding defense funding after several failed flight tests. Only since late 2021 has the Defense Advanced Research Projects Agency's (DARPA) Hypersonic Air-breathing Weapon Concept (HAWC, pronounced "hawk," also known as the "Hypersonic Air-breathing Conventional Munition," HACM; HCM) designed by Raytheon and Lockheed Martin passed full-scale flight tests, of which Lockheed Martin's design set a new record in March 2022 for sustained hypersonic flight. The weapon flew for 327 seconds at 65,000 ft for 300 miles.[26] In a parallel joint effort, the Army and Navy are developing a "surface-to-surface" long-range MHW capability which will yield 18 U.S. Army "DARK EAGLE" weapons (HGV) valued at $7B and 240 U.S. Navy Conventional Prompt Strike (HGV, also known as the "Conventional - Hypersonic Glide Body;" C-HGB) missiles valued at $21.5B.[27] These weapon systems will be fielded by 2023 and 2025, respectively, as part of the Pentagon's effort to speed the development of MHWs.[28] In total, the Pentagon's request for $3.8B in hypersonic development spending for five programs across all services in FY22 is a clear sign of the importance of this technology.[29] While the U.S. re-

mains late to need in this renewed effort to develop MHW technology, China, Russia, and recently North Korea have touted the development, successful testing, and operational deployment of their MHWs, as shown in Table 1. These technological leaps ahead of the United States by China and Russia offer a fresh opportunity to consider a modern nuclear deterrence posture to address the future nuclear stability paradigm in the liberal, rules-based, international world order.

Theory &Analysis

"Deterrence is the art of producing in the mind of the enemy ... the <u>fear</u> to attack."

—Dr. Strangelove

Deterrence Theory

In Latin, "deterrence" means "to frighten from or away."[30] As such, it works by retraining or preventing an adversary from acting because of perceived danger or trouble.[31] Thus, to be effective, deterrence is the "manipulation of an adversary's estimation of the cost/benefit calculation of taking a given action ... thereby convincing the opponent to avoid taking that action."[32] Joint Publication 1, *Doctrine for the Armed Forces of the United States*, summarizes "deterrence" as "a state of mind brought about by the existence of a credible threat of unacceptable counteraction."[33] In considering various forms of deterrence, Thomas Schelling describes deterrence as a "threat that leaves something to chance."[34] The uncertainty of a deterrent threat allows for a competition in risk-taking by adversaries as they both may choose to manipulate the risk in a crisis to accomplish their own political goals.[35]

First, to effectively apply a deterrent threat in a strategic environment, the deterrer may choose to deter by denial or punishment. Deterrence by denial leverages doubt that a particular action by the adversary will succeed by placing obstacles or threatening force which would inhibit the action.[36] This form of deterrence focuses on coercion and makes success doubtful by driving up the cost of the operation.[37] It denies the adversary the ability to achieve their desired goals through preferred or available means. Likewise, deterrence by punishment raises the perception of the resultant costs above what the adversary views as an acceptable, expected cost for the behavior.[38] In this form of deterrence, the deterrer holds something the aggressor values (e.g., civilian population) at risk as a powerful incentive to make another choice.[39] Second, deterrence must be considered credible in order for it to have its desired influence on the adversary. Credibility, in this instance, is a combination of the demonstrated capability to accomplish the communicated threat and communicated will of the deterrer to act on the threat. Lawrence Freedman frames the question as, "Does the enemy believe that the threats would be enforced?"[40] Finally, the deterrer must understand the nature of escalation in a deterrence relationship.

Herman Kahn defines this concept of escalation as "a competition in risk-taking, or resolve and matching local resources, in some form of limited conflict between two sides."[41] Mueller expounds on this by identifying that the "first goal" of escalation is to "make the target choose not to attack even though it has the ability to do so."[42] Likewise, Schelling argues that the nature of escalation increases the shared risk of escalation with each action and reaction by the adversaries.[43] While escalation is naturally understood to ascend vertically (i.e., increasing intensity or lethality), escalation can also occur horizontally (i.e., additional targets in a theater or expanding conflict to additional theaters and domains).[44] To better understand the nature of escalation, Kahn describes vertical escalation as a metaphorical "escalation ladder."

On this ladder, Kahn illustrates how escalation can incrementally rise along each of the 44 rungs he identifies.[45] These 44 rungs offer a conduit to exercise deterrence at any level while confining the enemy to a level far below its capability.[46] Consequently, Bernard Brodie's conceptualization of escalation aids Kahn's perspective on escalation theory by establishing clear thresholds by which a deterrer can set clear precedents in how they articulate their escalation strategy.[47] In doing so, Larsen and Kartchner describe how escalation dynamics can cause intense pressure for hesitancy or restraint in a conflict, further strengthening the nature of deterrence.[48] Figure 2 below depicts Kahn's famous escalation ladder.

AN ESCALATION LADDER

A Generalized (or Abstract) Scenario

———————————————— AFTERMATHS ————————————————

CIVILIAN
CENTRAL
WARS
- 44. Spasm or Insensate War
- 43. Some Other Kinds of Controlled General War
- 42. Civilian Devastation Attack
- 41. Augmented Disarming Attack
- 40. Countervalue Salvo
- 39. Slow-Motion Countercity War

(CITY TARGETING THRESHOLD)

MILITARY
CENTRAL
WARS
- 38. Unmodified Counterforce Attack
- 37. Counterforce-with-Avoidance Attack
- 36. Constrained Disarming Attack
- 35. Constrained Force-Reduction Salvo
- 34. Slow-Motion Counterforce War
- 33. Slow-Motion Counter-"Property" War
- 32. Formal Declaration of "General" War

(CENTRAL WAR THRESHOLD)

EXEMPLARY
CENTRAL
ATTACKS
- 31. Reciprocal Reprisals
- 30. Complete Evacuation (Approximately 95 per cent)
- 29. Exemplary Attacks on Population
- 28. Exemplary Attacks Against Property
- 27. Exemplary Attack on Military
- 26. Demonstration Attack on Zone of Interior

(CENTRAL SANCTUARY THRESHOLD)

BIZARRE
CRISES
- 25. Evacuation (Approximately 70 per cent)
- 24. Unusual, Provocative, and Significant Countermeasures
- 23. Local Nuclear War—Military
- 22. Declaration of Limited Nuclear War
- 21. Local Nuclear War—Exemplary

(NO NUCLEAR USE THRESHOLD)

INTENSE
CRISES
- 20. "Peaceful" World-Wide Embargo or Blockade
- 19. "Justifiable" Counterforce Attack
- 18. Spectacular Show or Demonstration of Force
- 17. Limited Evacuation (Approximately 20 per cent)
- 16. Nuclear "Ultimatums"
- 15. Barely Nuclear War
- 14. Declaration of Limited Conventional War
- 13. Large Compound Escalation
- 12. Large Conventional War (or Actions)
- 11. Super-Ready Status
- 10. Provocative Breaking Off of Diplomatic Relations

(NUCLEAR WAR IS UNTHINKABLE THRESHOLD)

TRADITIONAL
CRISES
- 9. Dramatic Military Confrontations
- 8. Harassing Acts of Violence
- 7. "Legal" Harassment—Retortions
- 6. Significant Mobilization
- 5. Show of Force
- 4. Hardening of Positions—Confrontation of Wills

(DON'T ROCK THE BOAT THRESHOLD)

SUBCRISIS
MANEUVER-
ING
- 3. Solemn and Formal Declarations
- 2. Political, Economic, and Diplomatic Gestures
- 1. Ostensible Crisis

———————————————— DISAGREEMENT—COLD WAR ————————————————

Figure 2. Kahn's Escalation Ladder.
(Reprinted from *On Escalation* by Herman Kahn, 1965, p. 39)

A key feature in Kahn and Schelling's escalation theories is the concept of "escalation dominance." Formally stated, "escalation dominance relies on superior brute force and war-winning strategy coupled with a credible threat" to communicate resolve in escalating to a level that the adversary will not.[49] Similarly, Larsen and Kartchner emphasize the importance of asymmetry paired with a credible ability to dominate conflict at any level of violence.[50] Because of this, in a future contingency, the United States might leverage an asymmetric nuclear capability to establish escalation dominance against a future adversary.[51] If used correctly, escalation dominance can eliminate the adversary's desire to escalate by raising the stakes enough to force capitulation instead of continued aggression. Thus, Kahn theorizes there is greater capability in a deterrence posture than a capability in being.[52] Moreover, to set the necessary conditions for escalation dominance, Kahn recommends a state achieve strategic superiority, secure a second-strike capability (e.g., SSBN or mobile ICBM), and provide relative safety for the civilian population.[53] MHWs can offer exquisite asymmetry and capability in posture, further strengthening a state's capacity for establishing escalation dominance at the outset of a conflict. However, in managing escalation, three critical considerations should be understood.

First, Larsen and Kartchner caution against inadvertent escalation based on miscalculation, miscommunication, or misunderstanding of enemy "red lines."[54] According to Miller, a state must foster the enemy's perception of the penalty for these errors to combat this possibility.[55] Second, Kahn offers that if a state chooses to de-escalate a conflict, there are three predominant avenues. These avenues are for a state to escape the costs of escalation, receive insurance from the adversary against the possibility of future escalation, or engage in bargaining with the adversary.[56] Third, Kahn acknowledges an unlikely possible paradox in escalation, specifically as it relates to nuclear conflict, that due to the fear caused by a nuclear attack, there might be no actual escalation between the adversaries.[57]

Understanding the Operational Environment: China

In analyzing the OE, the results provide an awareness of the strategic environment, defined as the "conditions, circumstances, and influences that [a]ffect national interest in and beyond the OE."[58] This analysis is critical to understanding relevant relationships and the interconnected nature of an adversary as a more extensive "system of systems" with factors that influence its behavior.[59] Thus, it is vital to understand China's perspective on strategic deterrence, Chinese aspirations and fears, and how China views the deployment of U.S. MHW technology.

Historically, China views itself as a regional hegemon with thousands of years of history and influence in the Indo-Pacific region. Neil Munro describes China's "geo-body" as constructed Chinese homeland territory in the region and instrumental in rejuvenating China's regional power.[60] This "geo-body" is defined as Tibet, Taiwan, Hong Kong, and Xinjiang.[61] As China persists in its domination of East Asia, Munro contends that China would not risk war over a territorial objective that falls outside of the "geo-body."[62] Therefore, as the sole territory outside of Chinese influence and control, Taiwan presents an essential objective for China to pursue. President Xi Jinping continues to tout a "China Dream" in 2049 as an impetus for reversing the "Century of Humiliation" suffered by China until 1949.[63] With this in mind, it is imperative to understand the views that shape China's nuclear deterrence rationale, which informs why China is pursuing MHWs in the face of a strategically superior United States and what the United States can do to establish an effective deterrence posture with China.

China's view of nuclear deterrence is fundamentally different from the United States' view of deterrence.[64] Specifically, China sees nuclear deterrence as defensive in nature, and they criticize the United States' "offensive" perspective in using nuclear weapons to establish deterrence.[65] The root of this perspective traces back to Chairman Mao Zedong, who believed China needed a small nuclear force to achieve "minimum deterrence" while avoiding the nuclear arms race between the United States and the Soviet Union.[66]

Viewed as part of the "third technological revolution" along with quantum computing and artificial intelligence, China is pursuing a new idea of "rapid response" within active defense (i.e., missile penetration capability) to deter the United States from threatening Chinese interests by having a means to defeat U.S. missile defense measures.[67] This "rapid response" capability also defends Chinese state interests and anticipated military conflict in the SCS and ECS.[68] MHW technology provides national prestige and security by building "strategic reassurance and mutual strategic restraint" in the Indo-Pacific.[69] Thus, the current assessment of Chinese hypersonic thinking is that China is not seeking parity with the United States but is not allowing itself to be limited to parity. In doing so, China's stated focus is pursuing performance over size in their nuclear requirements.[70] Because of this, China is focusing on building credible escalation management capabilities in place of a posture of strategic equivalence.[71] When paired with the recent Chinese FOBS demonstration, this affirms the Chairman's belief that China desires to achieve parity with the United States in the Indo-Pacific region by 2027 and parity, or even superiority, with the United States military by 2040.[72]

In developing MHWs for its use, China views the critical benefit of extreme speed, maneuverability, intercontinental range, precision, and penetration survivability as critical capabilities that can ensure the Chinese strategic nuclear deterrent while also delivering conventional strikes against heavily defended high-value targets.[73] Ironically, China appears to view dual-capable MHW technology as an opportunity to enable asymmetric escalation in a conflict with the United States, yet they do not consider escalation something that can be controlled in a nuclear war.[74]

Thus, China sees adopting innovative military technology, such as MHWs, as part of a broader effort to increase its hegemony in the Indo-Pacific while simultaneously restabilizing the strategic balance between itself and the United States.[75] This regional expansion and effort to stabilize their relationship with the U.S. might be why China is pursuing expanding its MHW arsenal by retrofitting DF-21 intermediate-range and DF-26 intercontinental-range ballistic missiles with HGV payloads for preemptive global strike capability.[76] In summary, China's nuclear strategy, historical fears, and perceived utility of MHWs in their conventional and nuclear forces presents the United States with a tremendous problem in light of a current inability to defend against such a threat. Thus, the United States requires a strategic deterrence posture that can credibly deter China from leveraging its asymmetric advantage to capture regional hegemony in the Indo-Pacific.

Defining the Problem, Identifying Assumptions, and Developing COAs

Secretary of the Air Force, Mr. Frank Kendall, recently said, "The United States needs deterrence that defeats aggression."[77] The United States must also maintain the ability to reestablish strategic deterrence should it fail. In doing so, a deterrence posture should provide the President two fundamental paths when attempting to restore deterrence: (1) impose a cost on the adversary they did not calculate (punishment) or (2) hold a vital target at risk the adversary did not

know the United States could hold at risk. Finally, the nature of establishing a deterrence posture should consider the impact of various assumptions.

Identifying Assumptions

The JPP methodology uses assumptions to discover how certain beliefs about the OE "enables the operational design to identify the areas of greatest risk to a mission."[78] Dr. Reilly describes the key characteristics of assumptions as "logical, realistic, and enable planning to continue."[79] Assumptions about U.S. strategy and capability in the analysis include the following. First, Air Force doctrine emphasizes rapid power projection from the United States has become our predominant defense strategy.[80] Second, long-range systems are essential to crisis management because they respond rapidly without the chaotic and slow mobilization of traditional land or naval forces.[81] Third, there is a prevalent assumption that new technology, such as MHWs or BMD, will allow great powers to escape from mutual deterrence (i.e., stalemate) with other nations that have a nuclear deterrent.[82] Finally, it is assumed that the operational deployment of such exquisite capabilities as MHWs offers the user the leverage needed to broker important power transitions regionally or in the international system.[83] While not comprehensive, critical assumptions that equally apply to the United States and China influence COA development.

In evaluating the effect of MHW technology in the United States and China, Dr. Acton notes that both nations appear to be developing the nascent weapons as an outgrowth of the broader technological surge in the field rather than a function of a strategic plan.[84] As such, as the Congressional Research Service notes, the tremendous growth of MHW technology is not a traditional numerical "arms race" like what was experienced during the Cold War but rather a competition in technology development.[85] Additionally, developing a military-specific application of MHW technology is assumed to be cost-prohibitive in large volume and, thus, only fielded in quantities for either state in the "tens" to "hundreds" category. A recent Sandia National Laboratory report estimated an HGV's "per unit" cost at $26-36M.[86] While comparable in size to the AGM-158 Joint Air-to-Surface Standoff Missile (JASSM), an HGV or HCM requires highly sophisticated design, heat-resistant construction, and materials that will exorbitantly drive production, maintenance, and supporting infrastructure costs to levels that cannot support more than a "niche" capability. Lastly, two factors are assumed to directly influence a nation's offensive/defensive advantage: technology and geography.[87]

COA Presentation

COA 1: Conventional Standoff Munitions

Theory: The nature of this posture focuses on speaking to enemy fears, identifying specific targets, and setting thresholds (i.e., "red lines") under which the United States will attack.[88] While China has a demonstrated FOBS HGV capability, the United States enjoys significant strategic superiority in all three legs of the nuclear triad. As such, this posture uses the current nuclear triad as a compilation of assured retaliation, strategic superiority, and a flexible, visible, and recallable force. This posture reflects Bernard Brodie's advice to establish a focus on "prevention of conflict" as its primary purpose.[89] Additionally, Brodie's perspective offers a posture of assured destruction enabled by the nuclear triad is sufficient for nuclear deterrence. However, as a counterbalance to this posture, former Secretary of State Henry Kissinger notes that there is a point at which additional strategic superiority in destructive power does not add additional strategic benefit.[90] Finally, in promoting deterrence, the United States must understand that deterrence stability is reinforced by degrading any confidence the adversary may have in an attack's coercive or strategic intent.[91]

Offensive Posture:

- Conventional standoff munitions such as the JASSM and Tomahawk Land Attack Missile (TLAM) provide long-range, moderate penetrating capability as conventional options below the nuclear threshold.

- ICBMs provide mutual vulnerability with Chinese mobile ICBMs.

- Submarine-launched Ballistic Missiles (SLBMs) provide assured retaliation to a large nuclear strike.

- Bombers (B-52/B-2) provide visibility, flexibility, dispersibility, and recallability.

Table 2. COA 1 Offensive Deterrence Posture Matrix (Response Options)

	US Indo-Pacific Regional Interest or Territory	CONUS/Hawaii or NC3 Systems
Adversary Conventional MHW Attack	JASSM/TLAM	(Strategic Ambiguity)
Adversary Nuclear MHW Attack	SLBM/B-52/B-2	ICBM/SLBM/B-52/B-2

Policy Recommendation:

- DoD would continue to fund the modernization of the nuclear triad with continued funding for defensive posture measures to bolster deterrence by denial strategy against the MHW threat from China.

- DoD would not continue to fund Hypersonic Weapons technology development and realign programmed money to defensive posture measures like the Hypersonic and Ballistic Tracking Space Sensor (HBTSS) and RQ-4 "Range Hawk" hypersonic sensors.[92]

- A dual-capable MHW attack on the Continental United States (CONUS), Hawaii, or Nuclear Command, Control, and Communications (NC3) capabilities (i.e., conventional or nuclear warhead) will be assessed as a nuclear attack on the homeland.

COA 2: Integrated Deterrence

Theory: The concept of "Integrated Deterrence" leverages a "whole of government" approach with all elements of the DIME Model (Diplomacy, Information, Military, and Economics).[93] The Interorganizational Cooperation doctrine published by the Joint Chiefs of Staff describes the "whole of government" approach as "integrates the collaborative efforts of USG departments and agencies to achieve unity of effort. Under unified action, a whole-of-government approach identifies combinations of USG capabilities and resources that could be directed toward the strategic objectives supporting U.S. regional goals as they align with global security priorities."[94] As a single strategy, "Integrated Deterrence applies deterrent force across all combatant commands, allies, partners, and services to create a deterrence message" to our adversaries.[95]

Secretary of Defense Austin defined Integrated Deterrence as "using existing capabilities and building new ones and deploying them all in new and networked ways. All tailored to a region's security landscape and in growing partnerships with our friends."[96] Secretary Austin further describes Integrated Deterrence as a way of networking tighter and innovating faster while modeling transparency and communication with other powerful nations.[97] Likewise, General VanHerck recently commented on the need for credible deterrence options below the nuclear threshold when deterring adversaries like China from attacking the homeland.[98] As the nascent theory continues to establish itself, "Integrated Deterrence" as a concept builds upon a couple of timeless deterrence principles. First, it is entirely based on the nature of "strategic ambiguity" where a deterrer will not provide specific details on what kind of response would be expected, yet clearly articulates the scope, scale, and the inevitable response of which the deterrer cannot be sure, nor could they be able to prevent. Second, Integrated Deterrence provides an opening for the idea of "escalation dominance" as previously noted. In escalating a crisis, the United States has the credibility, capability, and the will to escalate, and it must discover new means of dominating escalation while China maintains its "entangled" conventional and nuclear force posture.

Third, in dominating escalation, the United States must have options that are "severe and highly punitive in one sense, yet as non-escalatory as possible in another."[99] This form of asymmetric posture can offer leaders options that are "significantly more helpful and useful than a sledgehammer approach."[100] Finally, Integrated Deterrence as an offensive posture must threaten an adversary's values and must be adaptable as the adversarial landscape and leadership changes.[101]

Offensive Posture:

- "Whole of Government" approach with all options available at the conventional and nuclear levels. This option includes all elements of the DIME model, which can be employed individually or in combination to achieve a compounded kinetic, non-kinetic, or combined effect.

Table 3. COA 2 Offensive Deterrence Posture Matrix (Response Options)

	US Indo-Pacific Regional Interest or Territory	CONUS/Hawaii or NC3 Systems
Adversary Conventional MHW Attack	Integrated Deterrence	Integrated Deterrence
Adversary Nuclear MHW Attack	SLBM/B-52/B-2	ICBM/SLBM/B-52/B-2

Policy Recommendation:

- The USG approach to China would offer cooperative Arms Control options such as limited MHW testing, sharing HGV and HCM technology, and focusing on demonstrating capability while utilizing confidence-building measures that aid mutual trust and understanding.

- The USG would leverage arms control cooperation to ban Anti-Satellite (ASAT) weapons.

- The USG would limit the proliferation of MHW technology to other countries by banning hypersonic weapon technology exportation and materials necessary to build them.

COA 3: Hypersonic Cruise Missile—Conventional and Nuclear

Theory: "Where there is mutual fear, men think twice before they make aggressions on one another."[102] In pursuing "parity" in weapon-class capability, the United States can offset China's strategic asymmetric advantage while allowing for mutual vulnerability to stabilize the strategic environment in the Indo-Pacific. Moreover, using parity with China on MHW capability would give the United States the freedom to operate vertically and horizontally on Kahn's escalation ladder.[103] This capacity to dominate in escalation will also foster continued strategic superiority with China, which would also continue to drive a stable strategic environment.[104] As the preeminent world leader, the United States must also consider the various implications on the world stage should it choose not to pursue MHW technology.[105] As it currently stands, for any asymmetric capability that enables preemption as a "first choice" in seizing the initiative, a nation

attempting to defend itself must consider that the adversary is likely targeting the most dangerous capabilities.[106] Thus, the United States can secure the strategic balance by establishing mutual deterrence and posturing with the ability to strike first in a conflict against an adversary's counterforce capabilities.[107] Henry Kissinger offers a meaningful caution, "technical parity does not equal strategic parity."[108] Likewise, the United States must be cautious in operational deployments of these weapons in the Indo-Pacific, as excessive numbers of conventional or nuclear variants may have implications for strategic deterrence by undermining strategic stability.[109] This response can be mitigated by only developing a "niche" capability and limiting deployments to strategic bases in the United States.

Offensive Posture:

- Conventional and nuclear versions of DARPA's HAWC HCM are aided conventionally by U.S. Navy's CPS. Fielding a conventional and nuclear variant of the HAWC would provide operational planning or deployment flexibility and enable escalation dominance while leveraging the traditional nuclear triad.

Table 4. COA 3 Offensive Deterrence Posture Matrix (Response Options)

	US Indo-Pacific Regional Interest or Territory	**CONUS/Hawaii or NC3 Systems**
Adversary Conventional MHW Attack	HCM-C (USAF)	CPS (USN)
Adversary Nuclear MHW Attack	SLBM/B-52/B-2	HCM-N (USAF)

Policy Recommendation:

- The DoD must fund a new dual-capable Hypersonic Cruise Missile. In doing so, the $399M programmed funds for ARRW can be reallocated to support DARPA's successful Hypersonic Air-Breathing Weapons Concept (HAWC) program with a revised dual-capable design to support conventional and nuclear deterrence options.[110] The Army's Long-Range Hypersonic Weapon (LRHW) program would also be canceled, with remaining funds transferred to support the Navy's Conventional Precision Global Strike program portion.

- The USG should also pursue a tri-lateral arms control initiative with China and Russia. Tri-lateral arms control would focus on nonproliferation efforts limiting the exportation of MHW technology through the Missile Technology Control Regime (MTCR). In concert with this nonproliferation focus, the United States could align efforts to ban Anti-Satellite missile tests and FOBS tests while setting limits on fielded quantities of MHWs. The arms control options may also form an opportunity to establish a "hotline" between the United States and China similar to the current "hotline" between the United States and Russia. Finally, as an incentive to disentangle conventional and nuclear systems, the arms control agreement may foster a desire to specifically disentangle conventional and nuclear MHWs as a "first step" in a broader effort to draw more evident lines and firebreaks between nonstrategic and strategic nuclear forces.

The table below provides a summary of the deterrence posture COAs presented above. These COAs will be evaluated against each other in the "Strengths and Weaknesses" analysis and independently in the "Potential Risk" analysis in this paper.

Table 5. Summary of COA Descriptions

Course of Action	Description
COA 1: Conventional Standoff Munitions	Conventional standoff missiles such as the JASSM or TLAM provide a capability for the United States to either hold targets at risk with significant weapon volume or threaten to impose costs in conflict to deter China from using MHWs against the United States. Focuses development on space-based and point-defense capabilities to defeat an enemy MHW strike. Backed by the nuclear triad, this option provides the President with preexisting kinetic capabilities.
COA 2: Integrated Deterrence	Integrated Deterrence as a "whole of government" capability enables the United States with the strategic deterrence leverage to sufficiently threaten individual or compounded costs that target an enemy's values and drive perceived costs above the potential benefit. The use of the entire DIME model, while backed with the nuclear triad and the possibility of bilateral arms control as a central policy objective, this option provides the President with an innovative approach to strategic deterrence with nearly unlimited kinetic, non-kinetic, or combined options.
COA 3: HCM-C/N	Developing and deploying a dual-capable Hypersonic Cruise Missile leverages mutual vulnerability and escalation dominance to mitigate the asymmetric benefit gained by China. The DoD would consolidate existing MHW programs into a dual-capable HCM program (HAWC) and the U.S. Navy's CPS. The President would pursue an opportunity to facilitate tri-lateral arms control of MHWs and MHW technology between Russia, China, and the United States. Simultaneously, the United States would pursue banning the proliferation of MHW technology and setting limits on fielded MHW weapons, ASAT launches, and further FOBS testing.

Defined Criteria and Risk for COA Analysis

The analysis and evaluation of the defined COAs above are measured against several key weighted criteria. These criteria serve as factors that influence each COA, and they are given a percentage of "weight" based on the predetermined importance of how the criteria influence the defined COAs. Criteria for use in this analysis must be applicable for all three COAs, and they must be the most significant factors for consideration to prevent a dilution of effect if the number of criteria was unbounded. The five most important criteria for evaluating the deterrence posture COAs are: U.S. Government Deterrence Values, Flexible Deterrence Options, Surprise, Defeat Enemy Centers of Gravity (COGs), and Credibility. These criteria are defined below in Table 6, along with a determined "weight" to describe the amount of influence each criterion will have.

Table 6. Defined Evaluation Criteria

Criteria	Definition
U.S. Government Deterrence Values	Values that the USG favors in the formulation and execution of the current nuclear deterrence posture. These values, common across presidential administrations, are based on accepted deterrence theory. They provide insight into the likelihood of COA adoption and the possibility of broader influence on national deterrence strategy. **Weight: 10% (Marginal Importance)**
Flexible Deterrence Options (FDO)	"FDOs are preplanned, deterrence-oriented actions tailored to signal to and influence an adversary's actions. They are established to deter actions before or during a crisis. FDOs may be used to prepare for future operations, recognizing that they may create a deterrent effect. They are most effective when combined with other elements of national power (i.e., DIME)."[141] **Weight: 20% (Moderate Importance)**
Surprise	The element of surprise provides a key strategic advantage to its user in promoting unexpected effects on an adversary while leveraging the nature of surprise on the adversary to drive towards a defensive focus. Additionally, surprise can be used to increase the effectiveness of a tool because it can attack an enemy when they are least prepared. **Weight: 10% (Marginal Importance)**
Defeat Enemy COGs	A Center of Gravity (COG) is "the source of power that provides moral or physical strength, freedom of action, or will to act."[142] Therefore, an effective deterrence strategy must significantly defeat an adversary's COGs if there is a chance of success in the conflict. **Weight: 20% (Moderate Importance)**
Credibility	Credibility requires capability and will (i.e., resolve) for a deterrent threat of denial or punishment to be effective. Moreover, it is the combination of the demonstrated capability of a tool, and the communicated will use the tool that creates the necessary credibility to foster deterrence against an adversary. **Weight: 40% (Significant Importance)**

"Strengths and Weaknesses" *COA Analysis*

1. USG Deterrence Values: 10%

   USG Deterrence Values are rooted in political values and informed by decades of deterrence theory. These values tend to remain consistent with slight variance on major deterrence theory applications, but in formulating meaningful deterrence postures, these values can shift towards a particular theorist such as Brodie, Schelling, Kahn, or Kroenig. In this way, each deterrence posture's value has been standard across presidential administrations since the Cold War's end and provides the greatest utility.

2. Flexible Deterrence Options: 20%

   Flexible deterrence options are the third most important criterion to evaluate deterrence postures. This is primarily because each posture must have a flexible nature that the USG can use to form deterrence threats or leverage influence for various possible adversaries. As conditions change, the United States may also need to adjust the scale, breadth, or depth of a deterrence threat that the postures are designed to support.

3. Surprise: 10%

   The element of surprise is the least influential in evaluating the strengths and weaknesses of the deterrence postures. While important, surprising the adversary does little to prevent deterrence failure though it may be necessary for a successful preemptive or preventive conflict. However, it is essential to note that the USG does not view the prospective benefits of preemption or prevention as reasonable at present. Thus, surprise is considered, but it does not warrant substantial emphasis.

4. Defeat Enemy COGs: 20%

   In accomplishing a meaningful strategy, each deterrence posture should be able to defeat enemy COGs to increase costs beyond the perceived adversary benefit of the unwanted behavior. Likewise, should deterrence fail, defeating enemy COGs can aid escalation dominance by continually imposing unacceptable costs to an adversary's COGs early in the conflict or holding an adversary's COGs at risk not previously assessed as vulnerable to cause substantial fear of loss should the conflict continue. Thus, this criterion is the second most vital factor in evaluating a deterrence posture's strength or weakness.

5. Credibility: 40%

   The credibility of a deterrence posture is essential to enable a strong deterrent. Without credibility, the preexisting deterrence environment or deterrent threat issued by the USG would be met with doubt by the target state. This doubt emanates from either doubting the USG's capability to act or willingness to commit (i.e., national resolve) should deterrence fail.

In order to evaluate the three deterrence postures based on the weighted criteria above, the JPP methodology offers a means of comparing and analyzing the COAs against important predetermined criteria with a "Strengths and Weaknesses" format. Below, Table 7 focuses explicitly on the strengths and weaknesses of each deterrence posture across the evaluation criteria selected from JP 5-0 and defined above. Of note, while each COA has strengths and weaknesses, some strengths and weaknesses are more consequential to establishing the conditions under which China would be deterred from using a MHW against the United States.

Table 7. 'Strengths and Weaknesses' Summary

	COA 1: Conventional Standoff Munitions		COA 2: Integrated Deterrence		COA 3: HCM-C/N	
	Strengths:	**Weaknesses:**	**Strengths:**	**Weaknesses:**	**Strengths:**	**Weaknesses:**
USG Deterrence Values	1) Secure Second Strike 2) Strategic superiority 3) Flexible/Visible force (Bombers/Ships)	1) Not strategically symmetric 2) No credible homeland defense for nonnuclear attack 3) Not expedient	1) Strategic Ambiguity 2) 'Integrated' Approach 3) Escalation Dominance	1) Requires agreement amongst participants 2) Ambiguous nature 3) Requires strategy at outset	1) Mutual vulnerability 2) Escalation Dominance 3) Rapid, precision strike	1) Relies on Tri-Lateral Arms Control with RS and CH 2) Escalation Dominance not guaranteed 3) Possible misperception
Flexible Deterrence Options	1) Triad offers tiered options 2) Bombers are flexible, visible, and recallable 3) Standoff munitions enable some amount of escalation control	1) Cannot directly respond in kind to MHW threat 2) Strategic ambiguity may force 'Red Line' response 3) Large gap between standoff munitions and nuclear triad	1) Wide variety of options and combinations 2) Scalable level of intensity 3) Strategic ambiguity exploits adversary fears of capabilities	1) Ambiguity of threatened capabilities may be misinterpreted 2) Escalation Dominance complicates de-escalation 3) Choosing I.D. strategy difficult without integrator	1) Enables 'tit-for-tat' mutual vulnerability 2) Bomber-focused delivery is fundamentally flexible or recallable 3) Dual-use weapon system provides nonclear option	1) Limited to only HCM and Triad options 2) Not as expeditious as alternative options 3) Requires overseas basing and support
Surprise	1) Low-observable standoff munitions delay detection 2) Active Defense diminishes adversary response ability 3) SSBN can patrol/launch in close proximity undetected	1) Modern SAM capability can engage standoff weapons 2) Insufficient penetration to achieve surprise on HVTs 3) Large force difficult to conceal from detection	1) Immediate to delayed effect 2) Can operate in unexpected domains 3) Fundamental to unknown combination and desired effect until used	1) Clandestine or innocuous onset of effects not noticed 2) Delay caused in executing I.D. could lose effect 3) If uncoordinated, I.D. effort could be spoiled	1) Speed of positioning bomber with weapon 2) Secluded HVTs exposed 3) Denies enemy ability to adjust posture prior to strike	1) Easily spoiled with satellite imagery during weapon load (not stealthy) 2) Weapon is still detectable (not stealthy) 3) Requires strategic warning to avoid misperception
Defeat Enemy COGs	1) Capable standoff munitions can strike front-line defenses 2) Nuclear EMP options 3) Triad can hold all COGs at risk if escalates to nuclear	1) Unable to conventionally strike all enemy COGs 2) High quantity of standoff munitions required 3) May hit entangeled targets	1) Capabilities not limited to kinetic strikes 2) Adversary unable to plan for predictable response 3) Able to put effects on non-traditional targets to achieve strategic paralysis	1) Capabilities may not have desired effect 2) Effects may not be immediately observable 3) Capabilities may have civilian impacts (CDE)	1) Penetrating capability of HCM holds HVTs at risk 2) HCM enables defense suppression for 'night 1' force 3) Triad holds entire nation at risk if nuclear deterrence fails	1) Focused on kinetic strikes 2) May require significant quantity of HCMs for hardened or redundant targets 3) HCM use against COGs may cause undesired escalation
Credibility	1) Triad remains cornerstone 2) JASSM/TLAM used in conventional regional crisis 3) Regional allies would support U.S. involvment	1) Policy of 'strategic ambiguity' for attack on homeland remains in doubt 2) Defense is not currently credible against MHWs	1) Communicated resolve strengthens deterrent threat 2) Range of nonnuclear options increases likelihood 3) Strategic communication must leave "all options on the table"	1) Undemonstrated capability weakens deterrent threat 2) Excessive rhetoric by political leaders may cause doubt in ability to employ 3) Can't hesitate or retreat in Escalation Dominance	1) Strong resolve for use in regional crises 2) Dual-use enables selectable option for strike on homeland 3) HAWC successful test lends to growing capability of HCM weapon technology	1) Fledgling HGV program shows low capability 2) Not credible to threaten nuclear attack for conventional attack on U.S. 3) Non-deployed HCM degrades view of capability

Using the "Strengths and Weaknesses" format in Table 7, each COA is rank-ordered below, along with justification for its assigned ranking for each criterion. As a result, Table 8 summarizes the rankings and multiplies the rank with the criterion's "weight." The results across all five criteria are then added to yield a "total" for each COA.

USG Deterrence Values

COA 1: The deterrence posture presented in this COA focuses on using fielded conventional standoff munitions, underpinned by strategic superiority and a secure second-strike capability while eliminating the incentive for the United States to continue the "arms race" in hypersonic weapons development. This posture also enables the United States to continue to fund the nuclear triad modernization efforts while exploring avenues to defend the homeland and U.S. interests from MHW threats. This posture does not present a symmetric response capability to offset China's technological advances in MHW weapons, nor does it provide a credible solution for addressing the potential for a conventional attack on the homeland. The deterrence posture of this COA continues the "status quo" strategic posture formed by the nuclear triad to deter adversaries while assuring U.S. allies and those that the United States has extended the umbrella of nuclear deterrence.

COA 2: The deterrence posture presented in this COA uses an "integrated" approach across the broadest spectrum of military services and government agencies to determine a central strategy and synchronized effort. It utilizes strategic ambiguity in its posture while leveraging the ability to dominate escalation should conflict occur. However, the COA may cause unintentional escalation should the United States intentionally or unintentionally attack some of China's entangled strategic forces. Second, the deterrence posture presented in COA 2 opens the strategic aperture to include all options, levers, and combinations of the instruments of national power and its alliances in a singular strategic effort. This COA requires a determined strategy and agreement amongst the key participants before applying resources in conflict.

COA 3: The deterrence posture presented in this COA uses developing and deploying a dual-capable HCM weapon system. This posture uses the concept

of mutual vulnerability and escalation dominance to counterbalance the strategic asymmetric advantage sought by China with their deployed MHWs. This COA's nature directly offsets Chinese MHW capability, while the dual-capable weapon system allows for necessary kinetic strike capability below the nuclear threshold that the other COAs do not provide in sufficient strength. Where this option shows possible weakness is in the reliance on a possible tri-lateral arms control agreement between Russia, China, and the United States (which may no longer be an option after February 2022). Moreover, it may engender a larger potential for misperception by the Chinese in developing and deploying HCMs to establish the deterrence posture.

Flexible Deterrence Options

COA 1: This COA leverages the nuclear triad to provide a robust, tiered series of flexible and scalable options should a nuclear conflict occur with China. Likewise, the conventional standoff munitions capability provides escalation control below the nuclear threshold. Yet a shift in U.S. focus may diminish the strategic ambiguity required in establishing a deterrent threat because of an international crisis, failing to respond when an adversary crosses a "red line," or an inability to "respond in kind" to a Chinese MHW threat.

COA 2: This integrated approach utilizes the compounding effect of multiple capabilities across a "whole of government" scope that is scalable to any crisis or conflict. Additionally, this posture uses the nature of strategic ambiguity to exploit an adversary's fears about U.S. capabilities. Unfortunately, while this option uses a "whole of government" approach, it does not offer a "response in kind" capability before crossing the nuclear threshold. However, it can be expanded to include the additional capabilities and resources of the United States' allies and partners, which provides a strategic advantage that China does not currently enjoy. Of note, this COA enables tremendous flexibility and capacity for deterrence, yet its effects may require significant time to achieve the desired effect or may not be immediately observable. This option offers scalable intensity against an adaptable adversary, which provides a capacity to operate in unexpected domains, but in doing so, this posture requires synchronized coordination and strategic communication to achieve the most potent deterrent effect.

COA 3: Developing and fielding a dual-capable HCM weapon system in this posture would enable the USG to foster a strategic environment of mutual vulnerability with China. Moreover, should hostilities ensue, the bomber-centric delivery capability provides inherent flexibility and recallability to aid escalation dominance. Additionally, the dual-capable nature of the weapon system offers options that span both conventional and nuclear spectrums of conflict. Key weaknesses with this option include operational latency to generate combat capability at bomber bases and necessary overseas basing and support requirements.

Surprise

COA 1: This posture provides the element of surprise to enemy forces, provided by low-observable munitions and submarines patrolling nearby without detection. Moreover, the funds saved by canceling MHW programs will enable "active defense" research and development efforts to expedite the United States' ability to detect and respond to a MHW attack. However, this posture is weak against a robust surface-to-air missile (SAM) defense network capable of engaging these conventional munitions. With that in mind, the large force required to penetrate and defeat an enemy defense posture would be easily detected and spoil any chance of surprise.

COA 2: Integrated Deterrence exploits the element of strategic surprise by operating in unexpected domains with unpredictable combinations of elements of power. In doing so, these effects can be employed immediately or with strategic delay to achieve a desired response from the adversary. However, a delay in communicating a deterrent threat could lose the desired efficacy, much like applying economic sanctions on the eve of hostilities. Moreover, if a posture of Integrated Deterrence is not thoroughly coordinated, any unsynchronized actions could result in spoiling the compounded effect against the adversary.

COA 3: This posture uses the speed and flexibility of an intercontinental bomber force to rapidly project power from the United States or within the Pacific

AOR. Because of this, and due to the nature of using an HCM instead of an HGV, the United States can avoid overflight of uninvolved nations, focus on regional employment, and leverage the level trajectory of a cruise missile to not confuse the adversary by appearing like a nuclear submarine-launched ballistic missile or an intercontinental ballistic missile. However, with persistent overhead satellite imagery, robust radar, satellite tracking systems, and pervasive social media, the element of surprise related to this COA may be spoiled while the weapons are loaded on the airplane or once the weapons are released from the aircraft.

Defeat Enemy COGs

COA 1: This posture leverages conventional standoff munitions to strike against front-line adversary defenses while utilizing nuclear strike and electromagnetic pulse (EMP) capabilities to achieve effects against enemy COGs should the conflict escalate beyond the nuclear threshold. However, this option requires a significant number of conventional standoff weapons because they do not have the survivability and penetration capability to hold HVTs at risk in the same manner as MHWs. Additionally, conventional standoff munitions could target entangled targets and elicit an unintended escalatory response. Finally, this option cannot strike all of China's COGs with only conventional weapons and could leave the United States incapable of containing the conflict below the nuclear threshold in a conflict.

COA 2: The nature of this posture enables its user with capabilities not limited solely to kinetic effects. Moreover, it complicates the adversary's ability to predict how the United States might respond to a particular escalatory action. Because of this, it allows the United States to place effects on non-traditional targets in a coordinated fashion to achieve strategic vulnerabilities that may not have been evident prior to the conflict. However, the counterbalance to these advantages is that the employed capabilities may not have the desired effect against enemy COGs. Along this line, because the effects may not be immediately observable and wide-sweeping, the effects created by the strategic nature of Integrated Deterrence may result in unacceptable civilian collateral damage.

COA 3: Dual-capable HCMs in this posture can penetrate enemy defenses to hold a significant portion of enemy COGs at risk. Likewise, the posture leverages the survivability and extreme range of dual-capable HCMs to hold distant targets that are out of reach of other strategic weapons. However, while this posture focuses primarily on kinetic options to respond in kind to Chinese aggression, it does not leverage the key capability of a "whole of government" approach as described by the deterrence posture in COA 2. As such, a significant volume of extremely expensive assets will likely be procured to support a low-density stockpile. Finally, should the United States employ a dual-capable HCM against an enemy COG, the "entangled" nature of Chinese strategic forces may provoke an escalated retaliation. In this instance, the United States must be prepared to threaten additional, unacceptable costs to their COGs with nuclear HCMs to promote reestablishing strategic deterrence.

Credibility

COA 1: This COA focuses on using conventional standoff munitions in regional conflict, backed by the strategic nuclear triad. This option's credibility requires both the technological capability and the regional support for U.S.-led strikes to cement U.S. resolve to respond should China cross a "red line." Additionally, the savings achieved by realigning the $3.8B DoD budget for hypersonic research and development program is balanced by the lack of a credible deterrent for a conventional MHW attack on the US homeland.[111] Moreover, while the nuclear triad remains the cornerstone of American national security, this COA does not offer a comparable offensive capability to Chinese MHWs, nor is there an ability to defend U.S. countervalue interests.

COA 2: This posture's key strength is the wide range of nonnuclear options to meet any threat and threaten costs on an adversary asymmetrically. Because of this, the posture requires a strong communications strategy to demonstrate both capabilities and will to strengthen the deterrence environment. As such, strategic communication must retain all options "on the table" to best facilitate the strategic ambiguity of Integrated Deterrence. Likewise, credibility established by this COA might have the opposite effect if the United States provides excessive communication or an inability to de-escalate in an escalation dominance environment because of the limitations of the options and combinations available for use. As such, this posture requires a delicate balance with little hesitation to remain credible in the face of a MHW threat.

COA 3: While this posture does not guarantee escalation dominance, it enables a credible conventional or nuclear option as a response in kind or engaging in "tit-for-tat" escalation control to foster bargaining to resolve the conflict on terms favorable to the United States. Moreover, this option focuses on regional U.S. interests while offering a credible response to a strike on the US homeland by posturing with conventional and nuclear capability to "respond in kind." In doing so, this COA advances U.S. deterrence credibility and perceived costs to enemy forces. However, the posture may require a significant volume of HCMs beyond a "niche" capability which could exceed political support and budgetary fiscal constraints.

This analysis in Table 8 provides an objective/subjective evaluation to determine the best COA without considering the effect of risk in applying the deterrence posture. The evaluation of the potential for risk across each posture in four major categories of risk is presented in the following sub-section.

Table 8. "Strengths and Weaknesses" Weighted Evaluation

	COA 1: Conventional Standoff Munitions			COA 2: Integrated Deterrence			COA 3: HCM-C/N		
	Rank Score	*Weight*	*Result*	*Rank Score*	*Weight*	*Result*	*Rank Score*	*Weight*	*Result*
USG Deterrence Values	3	0.10	**0.3**	2	0.10	**0.2**	1	0.10	**0.1**
Flexible Deterrence Options	3	0.20	**0.6**	1	0.20	**0.2**	2	0.20	**0.4**
Surprise	3	0.10	**0.3**	1	0.10	**0.1**	2	0.10	**0.2**
Defeat Enemy COGs	3	0.20	**0.6**	1	0.20	**0.2**	2	0.20	**0.4**
Credibility	2	0.40	**0.8**	3	0.40	**1.2**	1	0.40	**0.4**
Total			**2.6**			**1.9**			**1.5**

Rank Score: 1 "Best" | 2 "Better" | 3 "Satisfactory"
(Rank Score) * (Weight) = Result (lower score indicates superior option)

COA Risk Analysis

In addition to the selected criteria listed above, the COA analysis assesses the importance of risk that could result from each deterrence posture. While there are several potential risks, this analysis focuses on the four potential risk areas defined in Table 9 below. These are the most significant risks that will impact the success or failure of a selected deterrence posture. As such, the risks are not "weighted" against each other until the final COA analysis as they are distinct and equally influential in the success or failure of a deterrence posture COA.

Table 9. Defined Areas of Potential Risk

Area of Potential Risk	Definition
Additional Excess Cost	Additional Excess Cost occurs when the development, application, or effect of a COA results in a cost of resources that surpasses the desired cost threshold or degrades the efficiency of the COA in order to achieve the desired effect.
Excess (Unacceptable) Collateral Damage	Excess (Unacceptable) Collateral Damage is an unpredicted increase in undesired collateral damage against an adversary that might cause the unintended escalation of hostilities in a conflict or an unfavorable international public opinion of the United States.
Deterrence Failure	Deterrence Failure occurs when the adversary's perception of the perceived benefit will exceed the perceived cost of a potential action/behavior after the United States has issued either a deterrence by denial threat or deterrence by punishment threat.
Escalation Dominance Failure	Failure of an escalation dominance strategy to control and defeat increasing escalation at every level after deterrence has failed is evident when the adversary continues to escalate the conflict vertically (intensity) or horizontally (breadth).

Presented below, each deterrence posture is evaluated based on the expected probability of occurrence and the severity of potential consequences. The COAs are scored independently of each other and are evaluated in Table 10 based on their average probability-consequence product across the four areas of potential risk.

Additional Excess Cost

COA 1: The probability is assessed as "improbable" (21-50%) because current conventional standoff munitions (e.g., JASSM, TLAM) are already developed and fielded as operational systems with little additional costs required for additional production and deployment. Moreover, the consequence factor is "minor harm" because of the relatively low per-unit cost compared to other more niche or exquisite weapon systems. The DoD would easily absorb any additional excess cost in the COA.

COA 2: The probability is assessed as "probable" (51-80%) because the nascent nature of Integrated Deterrence may foster unexpected cost increases in developing the concept of applying the deterrence posture as a "whole of government" model. Interestingly, the consequence of the COA is "minor harm" because of the flexible nature of the concept and its ability to adapt rapidly to the strategic deterrence environment.

COA 3: The probability is assessed as "probably" (51-80%) because of the significant research and development costs required to design, test, and field a low-density, highly complex weapon system. As expected, with such a high per-unit cost with little flexibility in programmed funding for any developmental setbacks, the consequence is assessed as "moderate harm" should further delays or program deviations occur.

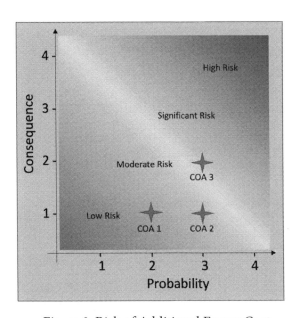

Figure 3. Risk of Additional Excess Cost

Excess (Unacceptable) Collateral Damage

COA 1: The probability is assessed as "highly unlikely" (0-20%) because of U.S. efforts to focus on counterforce targeting and the surgical precision of conventional standoff munitions. However, should unacceptable collateral damage occur, the consequence is assessed as "major harm" because of significant U.S. public sensitivity and the potential for an unintended Chinese escalated response.

COA 2: The probability is assessed as "probable" (51-80%) because of the potential for significant second and third-order effects caused by national instruments of power. These effects may include unintended impacts on civilian populations in China, such as food shortages or economic collapse. As such, the consequence of this criterion is rated as "major harm" because of the potential for widespread impacts on civilian populations.

COA 3: The probability is assessed as "improbable" (21-50%) because of the potential for collateral damage caused by the weapon failing in flight due to the extreme speed or not receiving an accurate position of itself during the time of flight due to the ionization of the plasma field surrounding the vehicle. Moreover, it is worth acknowledging that in the event of a conflict where the United States were to use a nuclear HCM against China, there would assuredly be collateral damage, yet it may not be considered "unacceptable" given the nature of the conflict and decision to employ such a devastating weapon. Finally, the rated consequence of such an occurrence is "extreme harm" because of the possibility of a nuclear HCM striking the wrong target or having sufficient yield to have excess collateral damage that was not intended.

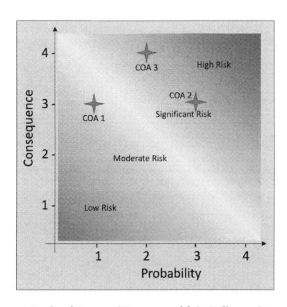

Figure 4. Risk of Excess (Unacceptable) Collateral Damage

Deterrence Failure

COA 1: The probability is assessed as "improbable" (21-50%) because of the lack of strategic parity or sufficient defense capability to offset the advantage of China's MHWs. Though U.S. conventional standoff munitions are highly capable weapons, the United States must strategically communicate its deterrent threats to foster an environment as Schelling would advise that "leaves something to chance." The consequences of deterrence failure are catastrophic and are assessed as "extreme harm" as deterrence failure would almost certainly result in a nuclear weapons exchange with China, which has the potential to be extremely costly to both countries.

COA 2: The probability is assessed as "highly unlikely" (0-20%) because of the emphasis on strategic ambiguity and the combined effects of any number of combinations of scope and scale of the elements of national power. However, the consequences of deterrence failure remain grim, and the assessment of "extreme harm" applies given the complexity of using Integrated Deterrence to deter; its ability to punish and control escalation is not immediate and carries some doubt about its effectiveness.

COA 3: The probability is assessed as "highly unlikely" (0-20%) because of the strength provided by mutual MHW vulnerability and the resultant environment of stalemate. While no exact "parity" between Chinese and United States MHWs, the capability for either side to successfully use an MHW is sufficient for "parity" in this research. The consequences of deterrence failing are also assessed as "extreme harm" because of the severe potential costs of a tit-for-tat escalation between the United States and China.

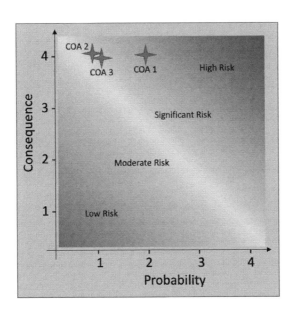

Figure 5. Risk of Deterrence Failure

Escalation Dominance Failure

COA 1: The probability is assessed as "probable" (51-80%) because of the Chinese perception of the United States' inability to credibly dominate escalation in a conflict with conventional standoff munitions across the Pacific Ocean. Moreover, should the conflict escalate beyond the nuclear threshold, the United States might succeed in escalation dominance, but it would require a significant escalation strategy with the nuclear triad to ensure success. This escalation strategy might result in extreme costs to the United States homeland, allies and partners, and deterrence capability with Russia or North Korea. As a result, the consequences of escalation dominance failure are assessed as "major harm" because of the likely costs imposed on the United States in escalating the conflict.

COA 2: The probability is assessed as "probable" (51-80%) because of the Chinese perception of the United States' inability to credibly dominate escalation due to the ambiguous and complex nature of Integrated Deterrence. Though Integrated Deterrence uses a "whole of government" approach, it likely cannot convey the compounded effect of multiple instruments of national power to establish escalation dominance at the outset of a conflict effectively. For this reason, the consequences of escalation dominance failure are assessed as "major harm" as in COA 1 because of the severe potential costs imposed by China if the United States were to attempt to dominate escalation and fail.

COA 3: The probability is assessed as "highly unlikely" (0-20%) because of the capacity of the United States to credibly threaten mutual vulnerability and dominance in escalation throughout a conflict. This is reinforced by the superior strategic advantage enjoyed by the United States in the present time, which may not continue well into the future. Nevertheless, while the likelihood of U.S. escalation dominance in a conflict with China in COA 3, there is always a chance that the strategy could fail. Because of this, the consequences assessed in escalation dominance failing are "major harm" as in the previous two COAs because of the significant potential damage and costs incurred by China on the United States.

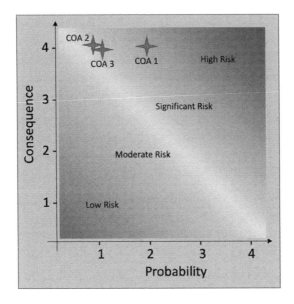

Figure 6. Risk of Escalation Dominance Failure

The following table (Table 10) shows the predicted probability and potential consequence severity of the four potential risk areas on each deterrence posture COA. The predicted probability is multiplied by the consequence sever-ity to determine a risk assessment for each posture in each potential risk area. Each of the four assessed scores is then averaged together for a deterrence posture to give an average risk assessment.

Table 10. Potential Risk Analysis Summary

	COA 1: Conventional Standoff Munitions			COA 2: Integrated Deterrence			COA 3: HCM-C/N		
	Probability	*Consequence*	*Assessment*	*Probability*	*Consequence*	*Assessment*	*Probability*	*Consequence*	*Assessment*
Additional Excess Cost	2	1	**2**	3	1	**3**	3	2	**6**
Excess (Unacceptable) Collateral Damage	1	3	**3**	3	3	**9**	2	4	**8**
Deterrence Failure	2	4	**8**	1	4	**4**	1	4	**4**
Escalation Dominance Failure	3	3	**9**	3	3	**9**	1	3	**3**
Average Risk			**5.5**			**6.25**			**5.25**

Probability: 4 "Very Likely" (81-100%) | 3 "Probable" (51-80%) | 2 "Improbable" (21-50%) | 1 "Highly Unlikely" (0-20%)
Consequence: 4 "Extreme Harm" | 3 "Major Harm" | 2 "Moderate Harm" | 1 "Minor Harm"
(Probability) * (Consequence) = Assessment (lower score indicates lower risk)   *JP 5-0, III 24-25*

The results of Table 10 above show how the potential risk of excess cost, excess (unacceptable) collateral damage estimate (CDE), failure of deterrence, or failure of escalation dominance affect each deterrence posture COA. While evaluating the probability of occurrence and severity of the consequence of these criteria is subjective, the quantitative rating and equal weighting allow for a comparative view of the combined potential risk for each deterrence posture. To this end, the deterrence posture that best minimizes the total potential and severity of risk is COA 3 (avg risk 5.25). COA 1 (avg risk 5.5) and COA 2 (avg risk 6.25) follow with higher amounts of average risk across the four criteria.

Findings

The combined results of the two analysis tables above showcase how COA 3 (HCM-C/N) exceeds the alternative COAs (see Table 11). However, though COA 2 (Integrat-

ed Deterrence) accumulated the highest average risk, as evidenced in the "strengths and weaknesses" analysis, it offers promise as an improved deterrence posture beyond COA 1 (Conventional Standoff Munitions).

These deterrence postures are fundamentally different from each other in their offensive orientation and policy recommendations. Nevertheless, they all share the same defensive posture due to the asymmetric characterization of how the People's Republic of China (PRC) might use a MHW against the United States or its vital interests. The "strengths and weaknesses" analysis above evaluates the various elements in favor of and against each posture for a particular criterion. However, assessing the overall "average risk" associated with each COA is crucial as it relates to the marginally understood view of Chinese deterrence theory. Table 11 applies a 70/30 weight ratio to the "strengths and weaknesses" and "potential risk" analyses to compute the total value to assess the COAs.

Table 11. Comprehensive Deterrence Posture Analysis

	COA 1: Conventional Standoff Munitions			COA 2: Integrated Deterrence			COA 3: HCM-C/N		
	Total Score	Analysis Weight	Result	Total Score	Analysis Weight	Result	Total Score	Analysis Weight	Result
Strengths & Weaknesses Analysis	2.6	0.70	**1.82**	1.9	0.70	**1.33**	1.5	0.70	**1.05**
Potential Risk Analysis	5.5	0.30	**1.65**	6.25	0.30	**1.875**	5.25	0.30	**1.575**
Total			**3.5**			**3.21**			**2.63**

(Total Score) * (Analysis Weight) = Result (lower score indicates superior option)

Finishing in third place, COA 1 (Conventional Standoff Munitions) focused on using long-range, low-observable, conventional standoff munitions, underpinned with the nuclear triad, to enable a deterrence environment with

China that would allow the DoD the opportunity to focus research and development spending on defense capability designed to defeat MHWs. Dr. Brad Roberts mentions that nuclear weapons "cast a shadow over the red zone of conflict by creating doubt about what price an adversary might pay for aggression in a way that defensive measures of the victim state do not."[112] However, Dr. Roberts also contends that the genuine concern is that nuclear deterrence may not retain credibility across the full spectrum of conflict.[113] Thus, as the "Strengths and Weaknesses" analysis discovered, the posture trailed COAs 2 and 3, but interestingly, this posture was assessed as a close second to COA 3 in the "Potential Risk" analysis. Likewise, in Table 11, the COA appears as a "moderate risk" alternative that relies on existing offensive capabilities to deter China from using a MHW against the United States. The Achilles' heel in this posture is the lack of credibility in retaliation in kind for a conventional strike on the homeland or naval task force while also presenting sufficient doubt in using conventional standoff munitions to strike at enemy COGs and the possibility of losing control of an Escalation Dominance strategy in conflict. COA 1 finished with the highest (i.e., worst) total score in the combined analysis, trailing the lead posture (COA 3) close to an entire point.

Finishing in second place, COA 2 (Integrated Deterrence) provided a novel approach to applying the DoD's perspective of using a "whole of government" approach to create a strategic deterrence environment with the ability and agility to simultaneously maximize the effect of the elements of national power. Of note, this posture reflects Ryan Henry's recommendation that "the United States must (1) have the means to determine what the enemy values, (2) determine what tools can hold those things at risk, and (3) an effective ability to communicate our ability to destroy those valued targets with our unique capabilities."[114] Said differently, the United States could adopt an Integrated Deterrence strategy that builds off its ability to credibly and capably impose negative costs without dramatically escalating the political stakes involved.[115] A vital benefit of this posture, an Integrated Deterrence strategy could use its indirect, asymmetric, and non-kinetic approach to rally more robust international support rather than quickly applying military force to complex problems.[116] Unsurprisingly, COA 2's use of Integrated Deterrence to form the foundation of the posture yields the highest average risk of the three postures because of its unique ambiguity and requirement to align all key decision-makers involved in the strategy collectively. As such, the risk inherent in Integrated Deterrence drives a tremendous benefit if China remains deterred while allowing for catastrophic failure should the United States expose an unsynchronized approach or any internal debate that China could exploit in seeking its strategic objective. Likewise, pursuing a bilateral MHW arms control agreement with China in this COA presents the most likely opportunity to establish mutual trust and restraint from escalatory behavior while inhibiting undesired proliferation. COA 2's "Strengths and Weaknesses" analysis demonstrated favorable results in a close second position because of the unique capabilities and effect of compounding multiple elements of power.

However, the "Potential Risk" analysis highlighted the significant risk of the posture's likelihood of causing excess CDE and loss of Escalation Dominance.

Finishing in first place, COA 3 (HCM-C/N) broke new ground in offering a dual-capable hypersonic cruise missile to establish a strategic deterrence environment built on mutual vulnerability, rapid precision strike, and Escalation Dominance. First, this posture leveraged a form of strategic parity in weapon class with an overall strategic superiority to drive towards a tri-lateral MHW arms control opportunity between the United States, China, and possibly Russia (if hostilities with Ukraine cease). Furthermore, should the United States successfully negotiate a tri-lateral MHW arms control treaty, the conditions will be ripe for adding China to a new tri-lateral strategic arms reduction treaty in 2026. Second, HCMs provide unique advantages in this posture over the more widely touted Chinese HGVs and FOBS capabilities by resolving many potential areas for misperception that HGVs could present, including overflying other adversarial nations, flight trajectories resembling ICBM or SLBM profiles, and the ambiguity of the target. Nevertheless, maneuverability, speed, penetration, survivability, and extreme range manifest a profound psychological effect on the target state's leaders by exposing them to the possibility of previously unlikely attacks.[117] Likewise, as Terry and Cone note, MHWs do not increase the first strike advantage for large and diverse arsenals such as the United States.[118] However, what dual-capable

HCMs can offer, according to Dr. Roberts, is the ability for a conventional HCM to form the basis of what would become a prenuclear deterrence system that amplifies the nuclear deterrence system.[119] Thirdly, as the "Strengths and Weaknesses" analysis showcased, a dual-capable HCM was the lead option of the three postures and provided the most credible capability to respond in kind to either a regional attack on U.S. territories and interests in addition to an intercontinental attack on the US homeland. In the "Potential Risk" analysis, COA 3's posture, underpinned by the nuclear triad, presented the lowest average risk because of its offensive, mutually vulnerable posture. The posture edged out COA 1 in the assessment but was not without concern as the likelihood for excess CDE was much higher than COA 1 and similar to COA 2. Finally, in the combined analysis, COA 3 presented the clear advantage as the most assertive posture with the least potential for risk compared to the other deterrence posture COAs. However, as capabilities mature and the international political landscape shifts, COA 3 may not remain the superior choice in ten or twenty years as the concept of Integrated Deterrence matures into a more credible and capable tool.

Conclusions

"Disruptive technology in the military domain should be viewed as instruments to assist humans in ensuring peace."

—Ajey Lele

The nearly overnight development and successfully demonstrated capability of Chinese MHWs has squarely put the United States in an uncomfortable and unfamiliar position in the Indo-Pacific and the broader world stage. At present, the United States remains unable to defend against, nor able to respond in kind to, a Chinese MHW attack either in the Indo-Pacific or at home in North America. Since the X-51 Waverider's only successful flight test, China and Russia have surpassed the United States in developing and fielding maneuverable hypersonic missile technology. s Russia remains embroiled in its ongoing war against Ukraine, China continues to focus on reasserting itself as a regional hegemon as it pursues the "China Dream" by 2049. As such, the United States must establish a deterrence posture to set the conditions under which China remains deterred from employing dual-capable MHWs against the United States and its interests.

With such a strategically complex problem in mind, this research greatly benefitted by using the Joint Planning Process to evaluate three possible courses of action against five important criteria (i.e., USG Deterrence Values, Flexible Deterrence Options, Surprise, Defeat Enemy COGs, and Credibility). These criteria were selected intentionally before COA development to assess each deterrence posture against these critical features. Additionally, the resultant COAs used elements of deterrence theory to formulate a potential deterrence posture with unique offensive capabilities and policy-based elements that could set the conditions for a strategic environment to foster the desired end state. These COAs were also assessed in a subsequent risk analysis to evaluate the probability, consequences, and combined potential for risk across four key criteria (i.e., additional excess cost, excess (unacceptable) collateral damage, deterrence failure, Escalation Dominance failure). Nevertheless, should the DoD implement a dual-capable HCM strategic deterrence posture as recommended by this research, and if Congress is to realign funds to support the program, necessary implications require consideration to fulfill the recommended deterrence posture.

Pursuing COA 3 and its implementation strategy should be considered for two main reasons before the operational deployment of the dual-capable HCM. First, funding for the HAWC program must provide essential research and development money to continue flight testing and develop a nuclear physics package. While reallocating the procurement funds for FY23 to research and development costs, the ARRW development program should be indefinitely postponed along with the Army's portion of the joint Army-Navy C-HGB program to consolidate as much as $7B in requested US Army funding to bolster DARPA's HAWC and the U.S.

Navy's CPS in the short term.[120] Though the Army will bristle at the request to cancel the LRHW program, the tested range of 1,725nm is more helpful in the European theater. However, it comes up short in the more urgent Pacific theater where territory is sparse, and the distance from Guam to Taiwan is more than 1,700nm, and from Guam to the Spratly and Paracel Islands is beyond 2,000nm.[121] Furthermore, the nearly $400M saved by shelving the ARRW program altogether will afford the National Nuclear Security Administration (NNSA) and USAF the program design funding to integrate the preexisting W80-4 Life Extension Program (LEP) warhead, valued at $11.2B, for use in both the Long-Range Standoff Weapon (LRSO) and the physics package of the nuclear variant of the HAWC HCM.[122] As former senior officials in the Defense Department from 2018-2020 suggest, the Long-Range Strike portfolio must be "affordably scaled to compelling numbers" to generate sufficient deterrence capability against Chinese aggression in the Western Pacific.[123] Such affordable scaling aligns with Lockheed Martin's Skunk Works' Vice President and General Manager, John Clark, who contends, "air-breathing hypersonic systems are a cost-effective solution to address the rapidly emerging threats in the global security arena."[124]

The second implication to consider is that research and development costs must continue to focus on investing in succeeding programs, given the current trailing position of the United States. In a 2021 GAO report, the DoD funded over 70 long-range strike programs from 2015 to 2024 totaling nearly $15B in defense spending.[125] By shaping the defense budget to a smaller number of MHW programs, the DoD can pursue developing future MHW requirements much like Porter and Griffin offer for long-range strike systems in the Indo-Pacific such as flying at Mach 17+, withstanding reentry conditions, refining the accuracy of a maneuverable reentry vehicle.[126] Likewise, George Leuenberger offers that MHWs principle utility will require rapid flight planning capability, navigation without GPS, sense-and-avoid self-detection, and mutually-cooperative functionality with other MHWs.[127] It is essential to weigh research and development costs against "per unit" production costs for such exquisite weapons systems. As Dr. Roberts points out, the United States must determine the correct number of HCM-C/Ns to produce to foster a "niche" MHW capability, which will serve as a deterrence hedge against the rapidly growing Chinese nuclear posture.[128] Therefore, given sufficient realignment of budget resources, setting appropriate requirements, and determining a production and sustainment volume based on a strategic plan, the United States could harness this emerging technology similarly to how the Rapid Capabilities Office (RCO) or DARPA pursues and refines new technology. While this posture is not a panacea for future nuclear deterrence, consideration of several anticipated counterarguments is necessary.

First, with the United States in such a disadvantageous position, proponents of the fledgling ARRW program may highlight the implicit value in pursuing both the ARRW and HAWC programs simultaneously as a hedge to ensure the success of one of the programs as the United States struggles to catch up to China and Russia. In doing so, Congress would continue to throw good money after bad. In a 2021 appropriations committee report, members noted their concern with ARRW's underwhelming progress. They raised an issue about the possibility of "concurrency" in the program, whereby the testing and correction of problems may occur during the initial production lot and generate additional costs.[129] Funding these expensive developmental programs in parallel will continue to limp the ARRW program while not sufficiently feeding a successful HAWC program with already limited fiscal resources. Thus, Congress must put our limited resources behind a succeeding program and re-evaluate the ARRW boost-glide program once HAWC is operationally fielded.

Second, many will contend that adding a new nuclear weapon to the U.S. strategic arsenal is untenable. While the 2011 New START treaty limited the number of deployed weapons to 1,550, it did not limit the type or number of weapons "counted" on a nuclear bomber, which was only counted as "1" under the treaty.[130] Furthermore, the 1993 Spratt-Furse Law prohibited the research and development of "tactical nuclear weapons," defined as nuclear weapons with an explosive power of less than 5 Kilotons.[131] In light of this, Keith Payne notes that a "no new nuclear weapons policy" is ill-suited

for our current realities and that the United States must be able to adapt its nuclear capabilities to address shifting adversary threats.[132] With this in mind, the current development of DARPA's HAWC program is cultivating a promising technology that drives an advantage for its user by exploiting technology that easily penetrates A2/AD environments and striking HVTs that are typically beyond the reach of most modern weapons.

Third, in the ongoing debate about continuing to fund the modernization of the nuclear triad, the proponents of eliminating the Long-Range Standoff (LRSO) nuclear cruise missile could use the opportunity to cancel the ARRW, HAWC, and LRSO programs simultaneously in order to pursue a dual-capable HGV for an ICBM or SLBM delivery. This option fails to consider that while the New START does not prohibit either side from deploying conventional warheads on ICBMs or SLBMs, the weapons would be counted under the limitations of the treaty regardless of their mated conventional or nuclear warhead, which may have an unforeseen impact on strategic stability due to issues of conventional-nuclear entanglement.[133] Additionally, the State Department previously stated that "there is no military utility in carrying nuclear-armed and conventionally-armed reentry vehicles on the same ICBMs or SLBMs."[134] As a result, Congress should fund a dual-capable HCM independently and in parallel with the LRSO program if the DoD is to construct the strategic deterrence posture selected in COA 3.

The significant attention garnered by MHWs by both government officials and news media gives further compulsion to the importance of additional research. Conducting future research on this subject matter will inevitably vary with interest, scope, and scale. However, there are several areas where additional research is essential. First, as previously mentioned, a study should be accomplished to evaluate how many of these exotic weapons the United States arsenal requires to successfully achieve a "niche" capacity while deterring China from using an MHW against the United States. The following question would be, how many weapons would the United States require to deter China and Russia successfully? Would these new HCM-C/N weapons also be blended into the current nuclear Operations Plans (OPLAN), and if so, what would be the appropriate ratio? Second, there is an ongoing debate about "long-range fires" in the DoD, and as a long-range weapon, MHWs are hotly contended by USAF, USN, and USA services for research and development funds. Military strategists and planners must write new doctrine that draws appropriate lines and defines necessary means for how these services might use their future MHWs in coordination. Third, researchers must conduct a thorough analysis to determine if the United States should continue to invest in HGV technology or if HCM technology is sufficient for deterrence with the technology's resources, capabilities, and utility as is currently planned. Likewise, in an already austere budgetary environment, these research and development choices are paramount because of the unusually high cost of the programs. Finally, studies must conduct and evaluate new, innovative methods for defeating MHWs. Defense strategy may favor "area denial" or "point defense" capabilities as hypersonic detection and tracking technology mature, and it is incumbent on researchers to offer strategies that offer the greatest utility for the least investment cost. For example, there may be a future opportunity in the U.S. Navy's SM-6 interceptor missile as a "point defense" capability in regional "hot zones" for MHW attacks in the terminal phase before striking the target. Unfortunately, this new missile's capabilities are unproven and still a decade or more from fielding.[135]

Ajey Lele said best, "[t]echnology may change warfare, but it doesn't determine warfare."[136] The advent of MHWs has ushered in a new class of weapons, but it has not fundamentally altered the principles of strategic deterrence. Underscoring the implications of China's 2021 hypersonic FOBS test, Admiral Richard testified to Congress that China's hypersonic FOBS weapon flew "the greatest distance and longest flight time of any land attack weapon system of any nation to date."[137] The impact of his testimony was reinforced with a stark reminder that "every other capability we have, rests on the assumption that strategic deterrence, and in particular nuclear deterrence, will hold."[138] Simultaneously, as the Biden administration crafts its Nuclear Posture Review goals, their priorities are straightforward: emphasize strategic stability, avoid arms racing, facilitate risk reduction, and promote arms control agreements.[139] The USG must use competition in fielding dual-capable MHW technology to guide how the State Department reassures and uses confidence-building measures within the mutual deterrence framework. Without such measures and dialogue to foster mutual trust, the United States and China have opportunities for misjudgment, misperception, fear, hostility, and deep mistrust.[140] Nevertheless, in a persistently cost-conscious budgetary environment, there must be no mistake that the cost of deterrence failure dwarfs the costs of maintaining deterrence, which indelibly underscores the unimaginable costs of deterrence failure. In summary, as the complex, rules-based, international world order continues to shift, the United States must be willing to invest in a strategic posture today that will enable the necessary deterrence for a peaceful tomorrow.

Endnotes

1   Colin Clark, "Aussies Unveil New Hypersonic Center Signal Distance from Ukraine Crisis," *Australian Defense*, 25 Jan 22, https://breakingdefense.com/2022/01/aussies-unveil-new-hypersonics-center-signal-distance-from-ukraine-crisis/?_ga=2.1717 13952.1867865877.1643161754-1340952986.1620763725.

2   Abraham Mahshie, "Hypersonics Defense," *Air Force Magazine*, 19 Jan 22, https://www.airforcemag.com/article/hypersonics-defense/.

3   Klon Kitchen, Interview with Klon Kitchen, The Dispatch Podcast, 22 Oct 21.

4   Kitchen, Interview with Klon Kitchen.

5   Richard Speier, "Hypersonic Missiles: A New Proliferation Challenge," *RAND*, 29 Mar 18, https://www.rand.org/blog/2018/03/hypersonic-missiles-a-new-proliferation-challenge.html.

6   Mahshie, "Hypersonics Defense."

7   Saunders and Logan, *China's Strategic Arsenal Worldview, Doctrine, and Systems* (S.l: GEORGETOWN UNIV PRESS, 2021).

8   Jason Sherman, "Hypersonic Weapons Can't Hide from New Eyes in Space," *Scientific American*, 18 Jan 22, https://www.scientificamerican.com/article/hypersonic-weapons-cant-hide-from-new-eyes-in-space/.

9   Jim Garamone, "Deterrence Ensures Great Power Competition," *U.S. Dept of Defense*, 7 Dec 21, https://www.defense.gov/News/News-Stories/Article/Article/2865253/deterrence-ensures-great-power-competition-doesnt-become-war-milley-says/.

10 Giulio Douhet and Dino Ferrari, *The Command of the Air*, Air University Press edition (Maxwell Air Force Base, Alabama: Air University Press, 2019), 24, 27.

11 Theresa Hitchens, "Joint US-Aus Hypersonic Cruise Missile Moves Ahead," *Breaking Defense*, 3 Sep 21, https://breakingdefense.com/2021/09/joint-us-australian-hypersonic-cruise-missile-moves-ahead/.

12 Travis Hallen and Michael Spencer, *Hypersonic Air Power*, Beyond the Planned Air Force (Canberra BC, ACT: Air Power Development Centre, 2018).

13 Demetri Sevastopulo and Kathrin Hille, "China Tests New Space Capability with Hypersonic Missile," *Financial Times*, n.d., https://www.ft.com/content/ba0a3cde-719b-4040-93cb-a486e1f843fb.

14 Roxana Tiron, "US Sees Rising Risk in 'Breathtaking' China Nuclear Expansion," *Bloomberg*, April 5, 2022, https://www.bloombergquint.com/politics/u-s-sees-rising-risk-in-breathtaking-china-nuclear-expansion.

15 Douhet and Ferrari, *The Command of the Air*, 16.

16 David Sanger and William Broad, "China's Weapon Tests Close to a 'Sputnik Moment,' US General Says," 3 Nov 21.

17 'Strategic Competition' is defined in the 2018 National Defense Strategy as the re-emergence of revisionist powers like Russia and China who desire to undermine and reshape the post-WWII rules-based, international world order.

18 Mahshie, "Hypersonics Defense."

19 Kroenig, *The Return of Great Power Rivalry: Democracy versus Autocracy from the Ancient World to the U.S. and China*.

20 Michelle Thompson, Ronny Reyes, and Chris Pleasance, "Retiring Vice Chairman of Joint Chiefs of Staff Reveals China Has Completed HUNDREDS of Hypersonic Missile Tests While US Has Done Just Nine and Warns That Beijing Is on Track to 'surpass' America," *Daily Mail*, 28 Oct 21, https://www.dailymail.co.uk/news/article-10142609/Vice-chairman-joint-chiefs-staff-warns-China-hundreds-hypersonic-missile-tests.html.

21 Mike Stone, "US in Hypersonic Weapon' arms Race' with China -Air Force Secretary," *Reuters*, 30 Nov 21, https://www.reuters.com/business/aerospace-defense/us-hypersonic-weapon-arms-race-with-china-air-force-secretary-2021-11-30/.

22 Pleasance, "Revealed: China 'has Tested TWO Hypersonic Orbital Capable of Breaching Missile Defences' as Panicked Analysts Say It 'Defies the Laws of Physics' and Is Weapon the US Has."

23 Ian Williams and Masao Dahlgren, "More than Missiles," *Center for Strategic and International Studies*, October 2019, https://www.csis.org/analysis/more-missiles-china-previews-its-new-way-war.

24 Yasmin Tadjdeh, "Hypersonic Missiles: Defense Department Accelerates Weapons Development," *National Defense*, Jul 19.

25 Thompson, Reyes, and Pleasance, "Retiring Vice Chairman of Joint Chiefs of Staff Reveals China Has Completed HUNDREDS of Hypersonic Missile Tests While US Has Done Just Nine and Warns That Beijing Is on Track to 'surpass' America."

26 Chris Ciaccia, "US Successfully Flight Tests Scramjet Powered Hypersonic Missile," *Daily Mail*, 27 Sep 21, https://www.dailymail.co.uk/sciencetech/article-10034359/U-S-successfully-flight-tests-Raytheon-hypersonic-weapon-Pentagon.html; Oren Liebermann, "US Tested Hypersonic Missile in Mid-March but Kept It Quiet to Avoid Escalating Tensions with Russia," *CNN Politics*, April 5, 2022, https://www.cnn.com/2022/04/04/politics/us-hypersonic-missile-test/index.html.

27 Anthony Capaccio, "Hypersonic Sticker Shock: US Weapons May Run $106 Million Each," *Bloomberg*, 12 Nov 21, https://www.bloomberg.com/news/articles/2021-11-12/hypersonic-sticker-shock-u-s-weapons-may-run-106-million-each.

28 Shaun Waterman, "US Army Likely to Field DOD's First Hypersonic Weapons in Next' Year or Two,'" *Air Force Magazine*, 16 Nov 21, https://www.airforcemag.com/us-army-likely-to-field-dods-first-hypersonic-weapons-in-a-year-or-two/.

29 Shannon Bugos, "Biden to Speed Development of Hypersonic Weapons," *Arms Control Today*, August 2021, https://www.armscontrol.org/act/2021-07/news/biden-speed-development-hypersonic-weapons.

30 Freedman, *Deterrence*, 7.

31 Freedman, *Deterrence*, 27.

32 Robert P. Haffa Jr, "The Future of Conventional Deterrence: Strategies for Great Power Competition," *Strategic Studies Quarterly : SSQ* 12, no. 4 (2018): 96.

33 *Doctrine for the Armed Forces of the United States*, Joint Publication 1-0 (Washington, DC: Joint Chiefs of Staff, 2017), I–11.

34 Jeffrey Arthur Larsen and Kerry M. Kartchner, eds., *On Limited Nuclear War in the 21st Century*, Stanford Security Studies (Stanford, California: Stanford Security Studies, an imprint of Stanford University Press, 2014), 148.

35 Francis J. Gavin, *Nuclear Statecraft: History and Strategy in America's Atomic Age*, Cornell Studies in Security Affairs (Ithaca: Cornell University Press, 2014), 148.

36 Karl P Mueller, "Conventional Deterrence Redux: Avoiding Great Power Conflict in the 21st Century," *Strategic Studies Quarterly : SSQ* 12, no. 4 (2018): 78.

37 Freedman, *Deterrence*, 37.

38 Mueller, "Conventional Deterrence Redux: Avoiding Great Power Conflict in the 21st Century," 78.

39 Freedman, *Deterrence*, 37.

40 Freedman, *Deterrence*, 36.

41 Kahn, *On Escalation*, 3.

42 Mueller, "Conventional Deterrence Redux: Avoiding Great Power Conflict in the 21st Century," 78.

43 Larsen and Kartchner, *On Limited Nuclear War in the 21st Century*, 149.

44 Catherine McArdle Kelleher and Peter J. Dombrowski, eds., *Regional Missile Defense from a Global Perspective* (Stanford, California: Stanford University Press, 2015), 244.

45 Kahn, *On Escalation*, 39.

46 Bernard Brodie, *Escalation and the Nuclear Option.* (Princeton, NJ: Princeton University Press, 1966), 88.

47 Kahn, *On Escalation*, 13.

48 Larsen and Kartchner, *On Limited Nuclear War in the 21st Century*, 152.

49 Larsen and Kartchner, *On Limited Nuclear War in the 21st Century*, 166.

50 Larsen and Kartchner, *On Limited Nuclear War in the 21st Century*, 157, 166.

51 Larsen and Kartchner, *On Limited Nuclear War in the 21st Century*, 165.

52 Kahn, *On Escalation*, 154.

53 Kahn, *On Escalation*, 161.

54 Larsen and Kartchner, *On Limited Nuclear War in the 21st Century*, 150.

55 Steven E. Miller, ed., *Strategy and Nuclear Deterrence*, Princeton Paperbacks (Princeton, NJ: Princeton University Press, 1984), xi.

56 Kahn, *On Escalation*, 233–37.

57 Kahn, *On Escalation*, 110.

58 *Joint Planning*, IV–4.

59 *Joint Planning*, IV–10.

60 Neil Munro, "China's Identity through a Historical Lens," 2022.

61 Munro, "China's Identity through a Historical Lens."

62 Munro, "China's Identity through a Historical Lens."

63 Lee, "Motivations Behind China's Hypersonic Weapons Development."

64 Liping Xia, "China's Nuclear Doctrine: Debates and Evolution," *Carnegie Endowment for International Peace*, 30 Jun 16.

65 Xia, "China's Nuclear Doctrine: Debates and Evolution."

66 Eleni Ekmektsioglou, "Hypersonic Weapons and Escalation Control in East Asia," 2015.

67 Xia, "China's Nuclear Doctrine: Debates and Evolution."

68 Zhanlue Xue, "Academy of Military Science: The Science of Military Strategy" (Beijing Military Science Press, 2013), 263.

69 Lee, "Motivations Behind China's Hypersonic Weapons Development"; Xia, "China's Nuclear Doctrine: Debates and Evolution."

70 Brad Radzinsky, "Chinese Views of the Changing Nuclear Balance," *War on the Rocks* (blog), 22 Oct 21, https://warontherocks.com/2021/10/chinese-views-of-the-changing-nuclear-balance/.

71 Radzinsky, "Chinese Views of the Changing Nuclear Balance."

72 Garamone, "Deterrence Ensures Great Power Competition."

73 Saunders and Logan, *China's Strategic Arsenal Worldview, Doctrine, and Systems.*

74 Larsen and Kartchner, *On Limited Nuclear War in the 21st Century*, 164; Tong Zhao, "Conventional Challenges to Strategic Stability: Chinese Perception of Hypersonic Technology and the Security Dilemma" (n.d.), https://carnegieendowment.org/2018/07/23/conventional-challenges-to-strategic-stability-chinese-perceptions-of-hypersonic-technology-and-security-dilemma-pub-76894.

75 Fravel, *Active Defense*, 24; Zhao, "Conventional Long-Range Strike Weapons of US Allies and China's Concerns of Strategic Instability."China has devised nine different military strategies, which the People's Liberation Army (PLA

76 Saalman, "Factoring Russian into the US-Chinese Equation on Hypersonic Glide Vehicles."

77 John Tirpak, "Kendall: Don't Mirror China on Hypersonics," *Air Force Magazine*, 19 Jan 22, https://www.airforcemag.com/kendall-dont-mirror-china-on-hypersonics-no-big-bang-on-abms/.

78 Reilly, *Operational Design: Distilling Clarity from Complexity for Decisive Action*, 54.

79 Reilly, *Operational Design: Distilling Clarity from Complexity for Decisive Action*, 56.

80 "Conventional Prompt Global Strike and Long-Range Ballistic Missiles:. CRS Report," Issues for Congress (Congressional Research Service, 2017).

81 Ekmektsioglou, "Hypersonic Weapons and Escalation Control in East Asia."

82 Patrick M. Morgan, T. V. Paul, and James J. Wirtz, "Complex Deterrence : Strategy in the Global Age," 2009, 7.

83 Morgan, Paul, and Wirtz, "Complex Deterrence : Strategy in the Global Age," 273.

84 Klare, "An 'Arms Race in Speed': Hypersonic Weapons and the Changing Calculus of Battle."

85 "Conventional Prompt Global Strike and Long-Range Ballistic Missiles:. CRS Report."

86 Abel Olguin, "Employment of Hypersonic Glide Vehicles - Proposed Criteria for Use," Project on Nuclear Issues 2014 Winter Conference (Sandia National Laboratories, 9-10 Dec 14).

87 Lele, *Disruptive Technologies for the Militaries and Security*, 16.

88 John Stone, "Conventional Deterrence and the Challenge of Credibility," *Contemporary Security Policy* 33, no. 1 (2012): 118, https://doi.org/10.1080/13523260.2012.659591.

89 Nathan B. Terry and Paige Price Cone, "Hypersonic Technology: An Evolution in Nuclear Weapons?," 2020.

90 Terry and Cone, "Hypersonic Technology: An Evolution in Nuclear Weapons?"

91 Keith B Payne, "Nuclear Deterrence in a New Age," *Comparative Strategy* 37, no. 1 (2018): 1–8, https://doi.org/10.1080/01495933.2018.1419708.

92 Kris Osborn, "The RQ-4 Global Hawk's New Mission- Hypersonic Weapons Testing," *The National Interest*, 24 Nov 21, https://nationalinterest.org/blog/buzz/rq-4-global-hawks-new-mission-hypersonic-weapons-testing-196825.

93 Glen VanHerck, "Threats to the Homeland" (Air Force Association Air Warfare Symposium, Orlando, FL, 3 Mar 21).

94 *Interorganizational Cooperation*, Joint Publication 3-08 (Washington, D.C: Joint Chiefs of Staff, 2017), II–2.

95 VanHerck, "Threats to the Homeland."

96 Jim Garamone, "Austin Discusses Need for Indo- Pacific Partnerships in the Future," *DOD News*, 27 Jul 21, https://www.defense.gov/News/News-Stories/Article/Article/2708315/austin-discusses-need-for-indo-pacific-partnerships-in-the-future/.

97 "Austin Calls for 'Integrated Deterrence' to Block China," *Inside the Pentagon*, 29 Jul 21.

98 VanHerck, "Threats to the Homeland."

99 Michael E. O'Hanlon, *The Senkaku Paradox: Risking Great Power War over Small Stakes* (Washington, DC: Brookings Institution Press, 2019), 149.

100 O'Hanlon, *The Senkaku Paradox*, 150.

101 Hans Binnendijk and James E. Goodby, "Transforming Nuclear Deterrence," *Managing and Transforming Nuclear Deterrence.*, 1997.

102 Lindsay and Gartzke, *Cross-Domain Deterrence*, 1.

103 Kahn, *On Escalation*, 137.

104 Kroenig, "Will Emerging Technology Cause Nuclear War?: Bringing Geopolitics Back In."

105 Joel. Larus, *Nuclear Weapons Safety and the Common Defense.* (Columbus: Ohio State University Press, 1967), 121.

106 Freedman, *Deterrence*, 87.

107 Brodie, *Strategy in the Missile Age*, 229–303.

108 Terry and Cone, "Hypersonic Technology: An Evolution in Nuclear Weapons?"

109 Brooks and Venable, "Chinese Hypersonic Weapons Developments Must Be Countered."

110 Bugos, "Biden to Speed Development of Hypersonic Weapons."

111 Bugos, "Biden to Speed Development of Hypersonic Weapons."

112 Roberts, *The Case for US Nuclear Weapons in the 21st Century*, 95.

113 Brad Roberts, "On Adapting Nuclear Deterrence to Reduce Nuclear Risk," *Daedalus (Cambridge, Mass.)* 149, no. 2 (2020): 82, https://doi.org/10.1162/daed_a_01790.

114 Dominick R. Pelligrini, ed., *Nuclear Weapons' Role in 21st Century US Policy*, Weapons of Mass Destruction Series (New York: Nova Science Publishers, 2010), 137.

115 Ekmektsioglou, "Hypersonic Weapons and Escalation Control in East Asia."

116 O'Hanlon, *The Senkaku Paradox*, 13.

117 Zhao, "Conventional Challenges to Strategic Stability: Chinese Perception of Hypersonic Technology and the Security Dilemma."

118 Terry and Cone, "Hypersonic Technology: An Evolution in Nuclear Weapons?"

119 Roberts, *The Case for US Nuclear Weapons in the 21st Century*, 135.

120 Greg Hadley, "Broken ARRW: Hypersonics Program Faces Uncertain Future after 2023," *Air Force Magazine*, March 28, 2022, https://www.airforcemag.com/arrw-hypersonics-program-faces-uncertain-future-after-2023/.

121 Lisa Porter and Michael Griffin, "Rethinking the Hypersonic Debate for Relevancy in the Pacific," March 23, 2022, https://breakingdefense.com/2022/03/rethinking-the-hypersonic-debate-for-relevancy-in-the-pacific/.

122 "W80-4 LIFE EXTENSION PROGRAM," Fact Sheet (National Nuclear Security Administration, January 2022), https://www.energy.gov/sites/default/files/2022-02/W80-4%20020122.pdf.

123 Porter and Griffin, "Rethinking the Hypersonic Debate for Relevancy in the Pacific," March 23, 2022.

124 "DARPA, AFRL, Lockheed Martin, and Aerojet Rocketdyne Team Successfully Demonstrate HAWC, Hypersonic Air-Breathing Weapon Concept" (Lockheed Martin Corporation, April 5, 2022), https://news.lockheedmartin.com/2022-04-05-DARPA,-AFRL,-Lockheed-Martin-and-Aerojet-Rocketdyne-Team-Successfully-Demonstrate-HAWC,-Hypersonic-Air-breathing-Weapon-Concept.

125 Liebermann, "US Tested Hypersonic Missile in Mid-March but Kept It Quiet to Avoid Escalating Tensions with Russia."

126 Porter and Griffin, "Rethinking the Hypersonic Debate for Relevancy in the Pacific," March 23, 2022.

127 George Leuenberger, "Hypersonics Overview" (SANDIA National Laboratory, 2019).

128 Roberts, *The Case for US Nuclear Weapons in the 21st Century*, 174.

129 Valerie Insinna, "House Appropriators Want to Shave $44M off Air Force's Flagship Hypersonic Program," *Air Force Magazine*, 12 Jul 21, https://www.defensenews.com/air/2021/07/12/house-appropriators-want-to-shave-44m-off-air-forces-flagship-hypersonic-program/.

130 Shannon Bugos, "New START at a Glance," Fact Sheet (Arms Control Association, February 2021), https://www.armscontrol.org/factsheets/NewSTART.

131 James Dao, "Senate Panel Votes to Lift Ban on Small Nuclear Arms," *The New York Times*, May 10, 2003, https://www.nytimes.com/2003/05/10/world/senate-panel-votes-to-lift-ban-on-small-nuclear-arms.html.

132 Payne, "Nuclear Deterrence in a New Age."

133 Bugos, "New START at a Glance."

134 Bugos, "New START at a Glance."

135 Mahshie, "Hypersonics Defense"; Sherman, "Hypersonic Weapons Can't Hide from New Eyes in Space."

136 Lele, *Disruptive Technologies for the Militaries and Security*, 32.

137 Tiron, "US Sees Rising Risk in 'Breathtaking' China Nuclear Expansion."

138 Tiron, "US Sees Rising Risk in 'Breathtaking' China Nuclear Expansion."

139 "Fact Sheet: 2022 Nuclear Posture Review and Missile Defense Review" (United States Department of Defense, March 28, 2022), https://media.defense.gov/2022/Mar/29/2002965339/-1/-1/1/FACT-SHEET-2022-NUCLEAR-POSTURE-REVIEW-AND-MISSILE-DEFENSE-REVIEW.PDF.

140 Xia, "China's Nuclear Doctrine: Debates and Evolution."

141 *Joint Planning*, E-1.

142   *Joint Planning*, GL-6.

Lt Col Dougherty is a graduate of Air University's School of Advanced Air and Space Studies (SAASS) at Maxwell Air Force Base, Alabama. Prior to attending SAASS, Lt Col Dougherty graduated from the School of Advanced Nuclear Deterrence Studies (SANDS) also located at Maxwell AFB. His SANDS thesis received particular distinction with two notable awards: the Commandant's Award for Research Excellence as the top overall research paper for the academic year and the Dean of Education's Joint Warfighting Award. Before attending SANDS and SAASS, he served as a Deputy Branch Chief, leading training and operations of two 32-member Contingency Command Crew teams and as a Nuclear Strike Advisor and Battle Staff Coordinator for the Commander of United States NORTHERN COMMAND. Lt Col Dougherty is an instructor and evaluator pilot with more than 3,300 flight hours, including 2,400 combat and combat support hours in the B-52H, MQ-9A, and MQ-1B. Lt Col Dougherty is currently assigned to the 20th Bomb Squadron at Barksdale AFB, Louisiana.

Doxfare as a Tool for Strategic Deterrence

Yang Liu

United States District Court for the District of Columbia

yl1085@georgetown.edu

ABSTRACT

The U.S. vision for an "open, interoperable, reliable, and secure internet" has been seriously challenged by state-initiated, -sponsored, or -encouraged hostile and destabilizing conduct in cyberspace. Before about 2018, the U.S.'s primary strategy was to passively defend and prevent such conduct. However, this strategy tended to be ineffective as the frequency and intensity of hostile activities in cyberspace substantially increased over time. Thus, the concept of defending forward was proposed by the Defense Department in 2018. And the need for a "fundamental rethinking" of cyberspace deterrence policy was highlighted by the Office of the Coordinator for Cyber Issues. In the following years, many new theories and approaches to cyberspace deterrence emerged. Building upon this trend of thoughts, this paper will propose a new and effective tool for deterring hostile state cyber operations – doxfare. This paper will first briefly overview the basics and recent developments of the deterrence theory in general and cyber deterrence theory in specific. It will then analyze why doxfare is an effective deterrence strategy. Lastly, it will address the ethical and legal concerns about doxfare.

Keywords: Doxfare, cybersecurity, cyberdefense, Strategic Deterrence

Doxfare como herramienta de disuasión estratégica

RESUMEN

La visión estadounidense de una "Internet abierta, interoperable, confiable y segura" se ha visto seriamente cuestionada por conductas hostiles y desestabilizadoras en el ciberespacio iniciadas, patrocinadas o alentadas por el Estado. Aproximadamente antes de 2018, la estrategia principal de Estados Unidos era defender pasivamente y prevenir tal conducta. Sin embargo, esta estrategia tendió a ser ineficaz ya que la frecuencia y la intensidad de las actividades hostiles en el ciberespacio aumentaron sustancialmente con el tiempo. Así, el concepto de defensa fue propuesto por el Departamento de Defensa

doi: 10.18278/gsis.8.1.4

en 2018. Y la Oficina del Coordinador de Cuestiones Cibernéticas destacó la necesidad de un "repensamiento fundamental" de la política de disuasión en el ciberespacio. En los años siguientes, surgieron muchas teorías y enfoques nuevos sobre la disuasión en el ciberespacio. Basándose en esta tendencia de pensamiento, este documento propondrá una herramienta nueva y eficaz para disuadir las operaciones cibernéticas estatales hostiles: el doxfare. En primer lugar, este artículo repasará brevemente los conceptos básicos y los desarrollos recientes de la teoría de la disuasión en general y de la teoría de la disuasión cibernética en particular. Luego analizará por qué la doxfare es una estrategia de disuasión eficaz. Por último, abordará las preocupaciones éticas y legales sobre doxfare.

Palabras clave: Doxfare, ciberseguridad, ciberdefensa, disuasión estratégica

Doxfare作为战略威慑工具

摘要

美国的"开放、可互操作、可靠和安全的互联网"愿景受到严重挑战，该挑战源于网络空间中由国家发起、赞助或鼓励的敌对和破坏稳定行为。大约在2018年以前，美国的主要策略是被动防御和预防此类行为。然而，随着网络空间中敌对活动的频率和强度逐渐大幅增加，这一策略往往是无效的。因此，美国国防部于2018年提出了"前沿防御"的概念。网络事务协调员办公室强调，需要对网络空间威慑政策进行"根本性反思"。在接下来的几年里，出现了许多新的网络空间威慑理论和方法。基于这一思想趋势，本文将提出一种新的有效工具来阻止敌对国家网络操作—— doxfare。本文首先就一般的威慑理论和具体的网络威慑理论的基础知识和最新发展进行简要概述。随后将分析为何doxfare是一种有效的威慑策略。最后，本文将分析有关doxfare的伦理及法律问题。

关键词：Doxfare，网络安全，网络防御，战略威慑

I. Introduction

The U.S. vision for an "open, interoperable, reliable, and secure internet" has been seriously challenged by state-initiated, -sponsored, or -encouraged hostile and destabilizing conduct in cyberspace. Before about 2018, the U.S.'s primary strategy was to passively defend and prevent such conduct. However, this strategy tended to be ineffective as the frequency and intensity of hostile activities in cyberspace substantially increased over time (Jaikaran, 2022). Thus, the concept of defending forward was proposed by the Defense Department in 2018 (U.S. Department of Defense). And the need for a "fundamental rethinking" of cyberspace deterrence policy was highlighted by the Office of the Coordinator for Cyber Issues (U.S. Office of the Coordinator for Cyber Issues, 2018). In the following years, many new theories and approaches to cyberspace deterrence emerged. Building upon this trend of thoughts, this paper will propose a new and effective tool for deterring hostile state cyber operations – doxfare. This paper will first briefly overview the basics and recent developments of the deterrence theory in general and cyber deterrence theory in specific. It will then analyze why doxfare is an effective deterrence strategy. Lastly, it will address the ethical and legal concerns about doxfare.

II. Analysis

1. *Theory of Cyber Deterrence*

The systematic study of deterrence began in the early cold war era, when scholars and policymakers were concerned about strategies to effectively deter adversaries from certain hostile behaviors through "the prospect of costs that outweigh the benefits" (Long, 2008). As summarized by Robert Jervis, till the conclusion of the cold war, there emerged three waves of developments in deterrence theory (Jervis, 1979). The first wave focused on the theoretical strategic implication of nuclear weapons, the second wave examined the possible bargaining tactics of mutual nuclear deterrence, and the third wave tried to rethink prior deterrence theories and reconcile them with empirical evidence.

After the collapse of the Soviet Union, the study of mutual deterrence between two symmetrical nuclear powers became less important, thus strategic balance and nuclear deterrence were no longer regarded as the most compelling issue. However, as the international security environment became more complex, a fourth wave of deterrence studies emerged which focused on the deterrence of "asymmetric threats" (Knopf, 2010). The early scholarship was particularly interested in the deterrence of terrorists and rogue states. While recently, cyber deterrence has increasingly become the most vital topic. An important characteristic of the fourth wave of deterrence studies is that the concept of deterrence has been substantially enlarged—whereas traditional scholars presumed that deterrence can only or at least be best achieved through military means, the fourth wave of scholarships revealed the value of non-military tools in deterring asymmetric threats.

The need for non-military deterrence tools is particularly compelling in the successful deterrence of hostile state activities in cyberspace. For example, the Cyberspace Solarium Commission, established pursuant to the John. S. McCain National Defense Authorization Act for Fiscal Year 2019, emphasized that successful cyber deterrence requires at least three strategies: (1) deterrence by norm shaping, (2) deterrence by denial, and (3) deterrence by punishment. Deterrence by norm shaping can potentially reduce the adversaries' incentive in conducting hostile cyber actions, while deterrence by denial, through increasing the resilience of critical networks, can reduce the benefits of hostile cyber activities (NATOCCDCOE, 2021). The Cyberspace Solarium Commission was particularly optimistic about the potential and value of these two strategies. However, according to the view of the Office of the Coordinator for Cyber Issues, these two strategies alone are not enough to deter determined and sophisticated adversaries like Russia, China, and Iran (U.S. Office of the Coordinator for Cyber Issues, 2018). Also, from a realistic perspective, a recent CRS report indicates that the success rate of these two strategies is relatively low at this time (Jaikaran, 2022).

Thus, it is important to also analyze possible strategies of deterrence by punishment and focus on ways by which to impose costs that are proportional to the harm suffered and substantial enough to deter adversaries. At the early stage of the study of cyberspace deterrence, many argued that deterrence by punishment is not effective in cyberspace. Traditional deterrence by punishment theory was developed during the cold war era and relies on several important premises like (1) the costs to develop, maintain, and use offensive capabilities are high, (2) non-state actors are largely irrelevant, (3) there are only a few of state actors to be deterred, (4) offensive measures are hardly clandestine, covert, or deniable (National Intelligence Council, 2021). Cyberspace, however, is largely characterized by "the inverse of" these conditions: the cost to launch an operation in cyberspace is relatively low, there are many potential hostile state, state-sponsored, or purely private actors, and cyberspace operations can be clandestine, covert, as well as deniable (Jaikaran, 2022).

That has been said, recent scholarships on cyberspace deterrence are increasingly reaching a consensus that while traditional, cold-war era strategies of deterrence by punishment no longer work well in cyberspace, deterrence by punishment in general is still a viable and important option. The different characteristic of cyberspace, however, requires a different strategy to impose costs and punish hostile actors in cyberspace. An important work in this area is the recent book by Michael Fischerkeller, Emily Goldman, and Richard Harknett on the "cyber persistence theory" (Fischerkeller, Goldman and Harknett, 2022). According to the cyber persistence theory, cyberspace is the third and new strategic domain that must be separated from the conventional and nuclear strategic environments. Traditional deterrence theory developed during the cold-war era focused on "escalation dominance" —being on a higher "escalation ladder" than its adversaries (Khan 1965). But such a strategy only works because there was a "common interest" in avoiding major nuclear war (Schelling 1966; Schelling 1960). In cyberspace, however, no such common interests exist because cyberspace is, on one hand, resilient on the macro level because no state will choose to simply cut off its connections to the internet for cybersecurity reasons, but on the other hand, vulnerable on the micro level because there are numerous opportunities to explore other's cyber vulnerability. Thus, the cyber persistent theory argued that the success is cyberspace deterrence is not escalation but prevailing on the same particular escalation level. To achieve such a goal, Fischerkeller, Goldman, and Harknett emphasized that hostile cyber operations should be deterred by cyber means, and the key is to exceed the adversities in the willingness and capacity to increase the frequency and intensity of cyber operations.

While this paper generally agrees with the cyber persistent theory, it argues that frequency and intensity are not the only factors to be considered in cyber deterrence. Instead, an effective strategy to deter hostile operations

in cyberspace shall also focus on the (1) diversity and (2) precision of the costs to be imposed. To begin with, by "diversity" of costs, I mean that it is not proper to restrict the methods and means of imposing costs to traditional cyber operations or cyber-attacks like the disruption of the adversaries' critical network. The reason is that offensive cyber capacities are scarce and fragile resources the development, deployment, and maintenance of which in adversaries' cyber systems can at the same time take a lot of time and resources and have substantial strategic values. Constantly using such resources as a response to all kinds of hostile foreign cyber operations risks losing many of such capacities because the effects caused by offensive cyber operations will inevitably lead the target state to have more information about the location, type, and mechanism of the malware or other offensive cyber tools that has been clandestinely deployed in their system. On the flip side of the coin, using offensive cyber measures as a tool of deterrence can also be unpersuasive sometimes. The reason is that states usually refrain from publicly discussing their cyber capacities and specifics of their cyber operations, as public discussion of such issues can also jeopardize their effective use of cyber capacities. Effective deterrence, however, usually requires the target state to be clearly informed of who is behind the attack and for what reason. Also, even where the identity of the attacker can be ascertained, the target state can only be effectively deterred if they clearly know beforehand about the consequences that will be incurred if they choose to conduct certain hostile cyber operations. Without publicly disclosing some of a state's cyber capacities, however, it is hard to inform its adversaries of its deterrence plan. And it is usually not enough to simply give a generally warn to a potential offender that a proportional cyber countermeasure will be launched against its hostile operations in cyberspace.

Thus, as already recognized by the Office of the Coordinator for Cyber Issues, it is important to develop "a broader menu of consequences that the United States can swiftly impose," which can not only make the deterrence more persuasive but also increase the policy and operational flexibility (U.S. Office of the Coordinator for Cyber Issues, 2018). Increasing the diversity of consequences, however, is not that easy in cyber deterrence because, as discussed above, successful cyber deterrence does not rely on escalation but on gaining advantages on a particular level of escalation. Thus, the critical issue is to find tools that can cause effects that are (1) different from that of a cyber-attack, (2) not severe enough to escalate the nature of competition/hostility, but (3) substantial enough to change the adversaries' calculus. As will be discussed infra, this paper argues that doxfare constitutes such an effective tool.

Besides the diversity of consequences, it is also important to increase the precision of the deterrence measures, which means that a state needs tools that are not indiscriminate and can only target the general public of another state but can selectively impose costs on a certain or certain group of actors. On one hand, the need to increase the precision directly relates to the diversity of malicious actors in cyberspace, as different actors are likely to have different motives, interests, and cost-benefits calculus. On the other hand, many hostile actors cannot be deterred by simply and indiscriminately imposing costs on the state they are living in. Sometimes this is because such indiscriminate countermeasure is not persuasive, for example, when a state has a close economic connection with another state, the latter's deterrence by economic sanctions will be less effective. Other times this is also caused by the discrepancy between the hostile actor's interest and the interest of the state the actor belongs to or the general public thereof. For example, when a state has a high corruption rate, the welfare of the general public does not necessarily relate to that of the elite or ruling class. Indiscriminate countermeasures usually can only impose costs on the former, but it is commonly the latter who makes the final decision to act hostilely in cyberspace. In the cold war era, the threats of a nuclear war mean that there are common interests within every society to prevent it, but with respect to today's grey zone competitions, it is increasingly vital to improving the capacities of precise deterrence.

2. Doxfare as An Effective Deterrence Tool

The above discussion has indicated that effective cyber deterrence neither relies on escalation nor simply cyber-attack countermeasures. Instead, diverse and precise deterrence tools are a must. This paper argues that doxfare can constitute such an effective tool. Doxfare can be defined as the "selective dissemination of true but confidential information" to influence the target state's internal or external affairs (Rodriguez, 2020). Doxing can be employed by both state and non-state actors for various reasons. For our discussion, doxfare is limited to those initiated by state actors in response to hostile cyber operations against actors responsible for such operations for the purpose of countermeasure and deterrence.

To begin with, doxfare is on the same or similar escalation level as most hostile cyber operations that are below the threshold of use of force/armed attack. Generally speaking, hostile operations in cyberspace can be classified into cyber operations and information operations. Doxfare itself is also essentially a kind of information operation, thus it is clearly of the same nature and can be a proportional response to information operations like Russian disinformation operation and election interference. Similarly, it is unlikely that doxfare will be on a higher escalation level than cyber operations like the SolarWinds Hack because the direct scale and effects of doxfare are relatively minimal compared to military cyber operations. Besides, doxfare can impose a different kind of cost on the target state than cyber operations. Whereas cyber operations can disrupt the target states' network and even cause physical impacts, doxfare can directly target the public opinion of actors responsible for their cyber operations. Moreover,

unlike cyber operations, a doxing operation can be executed in a particularly precise way. Usually, high rank political and military leaders or owners of corporations having a strong affiliation with their government are behind and responsible for sophisticated operations in cyberspace. A state can thus employ doxfare only to reveal true but confidential information about these people without influencing other people.

Since most hostile cyber adversaries have an authoritarian or totalitarian regime the legitimacy of which usually relies heavily on a public perspective of their high-rank leaders as being competent and righteous, which is usually false or exaggeration, doxfare can constitute a true threat to the legitimacy of the target state's leaders and/or its regime. Several examples can illustrate this point. The first example is the CIA's publication of Khrushchev's denunciation of Stalin in the closed session of the 20th Congress of the Communist Party of the Soviet Union (Encyclopaedia Britannica). In the speech, Khrushchev criticized Stalin for abusing powers, committing large-scale atrocities, failing to effectively prepare a national defense before the German invasion, and failing to properly manage World War II. Whereas there are some controversies on whether Khrushchev intended his speech to be secret or go public (Mayzel, 2013), the CIA played an important role in publicizing it around the world (Cull, 2006). The publication of the speech tended to be impactful in harming the reputation and the public perception of Stalin and Stalinism and in facilitating the De-Stalinization movement. Moreover, it also influenced to the 1956 Polish and Hungarian revolutions and "weaken[ed] the Soviet Union's control over the Soviet bloc" (Encyclopaedia Britannica).

Another example relates to the Hong Kong protests. Whereas it is not about state-initiated doxfare, it also illustrated its effectiveness of it against authoritarian regimes. Even before the outbreak of the 2019-2020 Hong Kong protests, doxing was already somewhat connected to social protests. Corruption and other public wrongdoing by government officials were usually the most important target in this period (Gao & Staner, 2013). The sensitivity of the government on such doxing of protestors was most clearly shown in the capture and sentencing of Gui Minhai, who was reportedly preparing to publish a book about the love life of Xi Jinping (Kuo, 2020). After the outbreak of large-scale protests, it was observed that doxing operations were widely employed against riot police officers as a method to protest the government and police's abuses of power (Ng, 2019). The protestors' doxing strategy tended to be effective, and the government was forced to impose a specific ban on the publication of personal details about police officers (The Guardian, 2019).

To be sure, it is not this paper's point that doxfare is a panacea. Recent empirical analysis of the doxing against Putin by Russian opposition activist Alexei Navalny seems to indicate that whereas doxing promoted civil engagement in public discussion and debates, it had minimal effects on the mobilization of Russians to conduct real-world

protests (Demydova, 2022). On the flip side of the coin, however, one might notice that the scope and degree of impacts of doxfare rendered it a proportional response to many kinds of hostile cyber operations and does not risk escalation. In any case, even if it might not be the proper tool in certain circumstances, it is definitely beneficial to add it to the cyber deterrence toolbox.

3. Ethical and Legal Considerations

The above discussion has explained how doxfare is an effective tool for cyberspace deterrence, this section will further discussion whether there are any ethical or legal problems with employing it. In fact, one reason why there are relatively few discussions on using information operations as a method of deterrence is less about its effectiveness and more about its ethicality and legality. For example, in Information Warfare and Deterrence, Gary Wheatley and Richard Hayes noted that "U.S. offensive information operation activities [are subject to two potential limits] (Wheatley and Hayes, 1996). First, media manipulation that involves government personnel providing false information is neither politically wise nor consistent with U.S. policy and law. Second, information attacks are attacks and therefore subject to international law." However, this section will conclude that unlike some other kinds of information operations, doxfare raises neither serious ethical nor legal challenges.

To begin with, an initial observation is that the ethical analysis of doxfare does not really relate to "media manipulation" because, unlike many other information operations, the definition of doxfare requires the underline information to be true. Thus, it is unlikely that doxfare can cause any serious domestic political or constitutional challenges,[1] and our analysis should focus on the properness of doxing as a method of grey zone competition.

We first consider the ethicality of doxfare. Currently, most scholarships on the morality of doxing only addressed private doxing, especially the morality of doxing for social justice or other political purposes and the challenges of digital vigilantism in connection with doxing, instead of state-initiated doxfare (Barry, 2021; Trottier, 2019). However, these discussions nevertheless provide a starting point for our discussion. One of the most important works in this field is David Douglas's "Doxing: a conceptual analysis" (Douglas, 2016). Douglas classified private doxing into three categories: (1) deanonymizing doxing ("establish[] the identity of a formerly anonymous or pseudonymous individual"), (2) targeting doxing ("reveal[] specific details of an individual's circumstances that are usually private, obscure, or obfuscated"), and (3) delegitimizing doxing ("reveal[] intimate personal information that damages the credibility of that individual"). State-initiated doxfare likewise can involve all three forms of doxing. For example, identifying the actor behind a particular cyber-attack can

be regarded as deanonymizing doxing, revealing the home address of a hostile foreign hacker is targeting doxing, and disclosing the history of corruption of the officer directing a cyber-attack is delegitimizing doxing.

In analyzing the morality of doxing, Douglas focused on motivation and necessity. He argued that private targeting doxing cannot be just because it "deliberately increases the risk of physical harm to the subject," while deanonymizing and delegitimizing doxing is permissible only if the motivation is "public interest" and only "to the extent necessary to reveal that wrongdoing has occurred." Whereas Douglas's analysis was focused on private doxing, I think it is equally applicable to state-initiated doxfare. Targeting doxfare is usually neither effective nor proper. For example, it is hard to see any deterrence value in telling the world where Putin and his family live in. On the flip side of the coin, if any of his family members are located overseas, revealing their private information of them can be regarded as a physical threat and thus risk escalating the nature of the competition. Deanonymizing and delegitimizing doxing, in contrast, are the essence of doxfare. By publicizing information about the identity of people responsible for hostile activities in cyberspace and their misbehaviors, deanonymizing and delegitimizing doxing can effectively impose costs on adversaries and thus achieve a deterrence effect. Like

---

1   There might still be some procedure issues like which agency should be primarily responsible for such operations as well as issues relating to the mechanisms of oversight. These, however, are out of the scope of this paper's discussion.

private doxing, in conducting doxfare, just cause, proper motivation, and proportionality is also key ethical requirements as well as the safeguard against misunderstanding on the adversaries' side and unnecessary escalation.

Next, we consider whether doxfare violates international law. Under international law, a state generally can take two self-help measures in response to another state's hostile or illegal activities that are below the threshold of use of force —retorsion, and countermeasure. The difference between retorsion and countermeasure is that whereas the former requires the self-help measures employed to be per se legal, the latter allows otherwise illegal measures to be used when certain conditions are met. Therefore, we need to answer two questions: (1) whether doxing another state violates international law, and (2) whether doxfare can still constitute a legitimate deterrent tool.

Under international law, there are generally three kinds of internationally wrongful acts that can be committed by one state against another – violation of the principle of non-intervention, illegal use of force, and armed attack except for self-defense purposes. Doxfare, at least deanonymizing and delegitimizing doxfare, clearly bears no relevance to the latter two rules. Thus, the key issue here is the principle of non-intervention, which is a well-established customary international law that prohibits states from coercively intervening in another state's affairs. According to the International Court of Justice, a prohibited intervention can be established by two elements: (1) the hostile

activity relates to the target state's domaine reserve (matter that that "each State is permitted, by the principle of State sovereignty, to decide freely"), and (2) the hostile activity involves methods "of coercion in regard to such choices" (Nicar. v. U.S.). Doxfare will likely meet the first element if the target is a high-rank public official as the purpose for such doxfare is usually to deter adversaries by revealing the misconducts of responsible actors to influence public opinion of them and even the political process and the legitimacy of their government or regime, which are within a state's domaine reserve. In contrast, it is unlikely that any kind of doxfare can be regarded as "coercive" because it does not deprive the free will of the target state to decide its politics and any other matters within its domaine reserve.

However, it is worth noting that whereas doxfare is unlikely to constitute an illegal intervention under the current rule, it is improper to conclude that it does not raise any international law concerns. First, the law of non-intervention might evolve over time or, whereas not very likely, in the near future. There are already some disagreements between states on how to apply the non-intervention principle to cyberspace. States including the Netherlands and Germany adopt the traditional view and argue that the coercion requirement can be met only if the cyber operation is "specifically designed to compel the victim State to change its behavior with respect to a matter within its domaine reserve" (Schmitt, 2019; The Federal Government of Germany, 2021). States including Australia and New Zealand,

in contrast, argue that the coercion requirement should be less strict in analyzing cyber activities and can be met
as long as the victim state's "ability to
control or govern" its domaine reserve
is influenced (The Federal Government
of Australia, 2017; New Zealand Foreign Affairs & Trade, 2020). Still, there
are scholars concerned with Russian
disinformation operations and election
interference who proposed to replace
the coercion test with a lowered "disruptive" test. If these approaches are adopted, the scope of prohibited intervention
will substantially enlarge, and doxfare
might raise legal issues. Second, even if
doxfare does not constitute an internationally wrongful act of a state against
another state, it might still be illegal if it
violates the human rights of the target.
Article 17 of the International Covenant
on Civil and Political Rights establish-
es a universal human right to privacy.
Some have argued that this rule applies
not only to the state's domestic activities
but also an infringement on the privacy
of other states' citizens (Sinha, 2013). If
Article 17 applies to doxfare, we need to
then consider whether the targets, who
can usually be political or public figures,
are subject to less strict protection and
whether the violation of their privacy
can be justified for national and international security reasons. These questions
are challenging and have yet to be sys-
tematically addressed, but they clearly
shed a shadow on the per se legality of
doxfare.

That has been said, even if some
doxfare can be regarded as illegal, the use
of it as a deterrence tool and response to
other states' hostile cyberspace activi-

ties is still likely a legal countermeasure.
The law of countermeasure is a cornerstone of international law's system of
self-regulation and guides a state's legal
peacetime response to another's internationally wrongful acts (Schmitt and
Watts 2021, 384; Crawford 2012, 585).
Unlike many ambiguous international
law concepts, the content of the law of
countermeasure has been clearly artic-
ulated by the U.N. (International Law
Commission, 2001). According to the
U.N. standard, a state is authorized to
respond to another state's illegal activity against it with an otherwise illegal
method if the purpose is to induce the
other state to comply with obligations
it violated and if there is a reasonable
chance of success. It is undoubtedly
possible and plausible for a state to design a legal doxfare strategy as an effective tool for deterrence.

III. Conclusion

This paper analyzed doxfare as
an effective tool for cyberspace
deterrence. It argued that doxfare is effective because it can impose
a cost on adversaries that is substantial
enough but not to a level that risks unnecessary escalation. Cyber deterrence
strategy needs to include doxfare in the
toolbox because it can increase the diversity and precision of consequences
that can be imposed. Whereas certain
kinds of doxfare, like the revelation of
the precise physical location of attackers, can be an improper move under
most circumstances, doxfare as a method of deterrence generally violates neither ethical nor legal principles.

Yang Liu holds a Juris Doctor degree from Georgetown University Law Center and is currently clerking at the United States District Court for the District of Columbia as a Criminal Law Fellow. His academic interests and primary area of research include the intersection between law, national and international security, emerging technologies, and individual rights. He welcomes opportunities for continued research and collaboration.

References

1. Barry, Peter. 2021. "Doxing Racists." *The Journal of Value Inquiry* 55: 457-474.

2. Case Concerning Military and Paramilitary Activities in and against Nicaragua (Nicaragua v. United States of America), 1986 I.C.J. 14.

3. Crawford, James. 2012. *Brownlie's Principles of Public International Law*. Oxford University Press.

4. Cull, Nicholas. 2006. *1956 All That … U. S. Public Diplomacy And Khrushchev's Secret Speech*. CPD Blog. https://uscpublicdiplomacy.org/blog/1956-all-u-s-public-diplomacy-and-khrushchev%E2%80%99s-secret-speech.

5. Demydova, Viktoriia. 2022. "Doxing as a Form of Online Activism: Case of Alexei Navalny's Film A Palace For Putin." *Turkish Studies* 16(3): 909-929.

6. Douglas, David. 2016. "Doxing: a conceptual analysis." *Ethics and Information Technology* 18: 199-210.

7. Encyclopaedia Britannica. "Khrushchev's secret speech." Accessed August 4, 2023. https://www.britannica.com/topic/Stalinism.

8. Fischerkeller, Michael, Emily O. Goldman and Richard J. Harknett. 2022. *Cyber Persistence Theory: Redefining National Security in Cyberspace*. Oxford University Press.

9. Gao, Li and James Staner. 2013. "Hunting corrupt officials online: the human flesh search engine and the search for justice in China." *Information, Communication & Society* 17(7): 814-829.

10. International Law Commission, Draft Articles on Responsibility of States for Internationally Wrongful Acts, 53rd Sess., G.A. Supp. No. 10 (A/56/10) (2001).

11. Jaikaran, Chris. 2022. *Cybersecurity: Deterrence Policy*. Congressional Research Service. https://crsreports.congress.gov/product/pdf/R/R47011.

12. Jervis, Robert. 1979. "Deterrence Theory Revisited." *World Politics* 31(2): 289-324.

13. Khan, Herman. 1965. On Escalation. London: Routledge.

14. Knopf, Jeffrey. 2010. "The Fourth Wave in Deterrence Research." *Contemporary Security Policy* 31(1): 1-33.

15. Kuo, Lily. 2020. *Hong Kong bookseller Gui Minhai jailed for 10 years in China.* The Guardian. https://www.theguardian.com/world/2020/feb/25/gui-minhai-detained-hong-kong-bookseller-jailed-for-10-years-in-china.

16. Long, Austin. 2008. *Deterrence — From Cold War to Long War.* RAND Corporation. https://www.rand.org/pubs/monographs/MG636.html.

17. Mayzel, Matitiahu. 2013. " Israeli Intelligence and the leakage of Khrushchev's 'Secret Speech.'" *Journal of Israel History* 32(2): 257-283.

18. National Intelligence Council. 2021. *Global Trends 2040: A More Contested World.* https://www.dni.gov/files/ODNI/documents/assessments/GlobalTrends_2040.pdf.

19. NATOCCDCOE. "The Process of the U.S. Cyberspace Solarium Commission - CyCon 2021." YouTube, June 12, 2021. https://www.youtube.com/watch?v=OBUy7aGNiCQ.

20. New Zealand Foreign Affairs & Trade. 2020. *The Application of International Law to State Activity in Cyberspace.* https://www.dpmc.govt.nz/sites/default/files/2020-12/The%20Application%20of%20International%20Law%20to%20State%20Activity%20in%20Cyberspace.pdf.

21. Ng, Brady. 2019. *Hong Kong protesters are naming and shaming police officers.* Nikkei Asia. https://asia.nikkei.com/Opinion/Hong-Kong-protesters-are-naming-and-shaming-police-officers.

22. Rodriguez, Manuel. 2020. *Disinformation Operations Aimed at (Democratic) Elections in the Context of Public International Law: The Conduct of the Internet Research Agency During the 2016 US Presidential Election.* Cambridge University Press.

23. Schelling, Thomas C. 1960. The Strategy of Conflict. Cambridge, MA: Harvard University Press.

24. Schelling, Thomas C. 1966. Arms and Influence. New Haven, CT: Yale University Press.

25. Schmitt, Michael N., and Sean Watts. 2021. "Collective Cyber Countermeasures?" *Harvard National Security Journal* 12: 374-410.

26. Schmitt, Michael. 2019. *The Netherlands Releases a Tour de Force on International Law in Cyberspace: Analysis.* Just Security. https://www.justsecurity.org /66562/the-netherlands-releases-a-tour-de-force-on-international-law-in-cy berspace-analysis/.

27. Sinha, G. 2013. "NSA Surveillance Since 9/11 And The Human Right To Privacy." *Loyola Law Review* 59: 861-946.

28. The Federal Government of Australia. 2017. *Australia's Position On The Application Of International Law To State Conduct In Cyberspace.* https://www.dfat. gov.au/sites/default/files/application-of-international-law-to-cyberspace.pdf.

29. The Federal Government of Germany. 2021. *On the Application of International Law in Cyberspace.* https://www.auswaertiges-amt.de/blob/2446304/32e7b2 498e10b74fb17204c54665bdf0/on-the-application-of-international- law-in-cyberspace-data.pdf.

30. The Guardian. 2019. *Hong Kong bans publication of police personal details, including photos.* https://www.theguardian.com/world/2019/oct/26/hong-kong- bans-publication-of-police-personal-details-including-photos.

31. Trottier, Daniel. 2019. "Denunciation and doxing: towards a conceptual model of digital vigilantism." *Global Crime* 21(3-4): 196-212.

32. U.S. Department of Defense. 2018. *Summary: Department of Defense Cyber Strategy.* https://media.defense.gov/2018/Sep/18/2002041658/-1/-1/1/CYBER _STRATEGY_SUMMARY_FINAL.PDF.

33. U.S. Office of the Coordinator for Cyber Issues. 2018. *Recommendations to the President on Deterring Adversaries and Better Protecting the American People from Cyber Threats.* https://www.state.gov/wp-content/uploads/2019/04/Rec ommendations-to-the-President-on-Deterring-Adversaries-and-Better-Pro tecting-the-American-People-From-Cyber-Threats.pdf.

34. Wheatley, Gary and Richard E. Hayes. 1996. *Information Warfare and Deterrence.* NDU Press Book.

Havana Syndrome: A Case of Fifth Generation Warfare

Armin Krishnan, PhD

East Carolina University

KrishnanA@ecu.edu

Abstract

Over a thousand individuals have been diagnosed with Havana Syndrome and many experts have concluded that it is caused by an unknown DEW. The article argues that Havana Syndrome has all the characteristics of a Fifth Generation Warfare (5GW) attack and that the 5GW theory may provide a suitable framework for analyzing the intent of attackers, which will help countering and deterring future attacks. 5GW posits that the manipulation of perception and narratives is critical to achieving an aggressor's objectives, that attacks are covert or ambiguous, that violence is highly dispersed, and that a major objective is to undermine the legitimacy of a state and its ability to function.

Keywords: Havana Syndrome, Directed Energy Weapons, Fifth Generation Warfare

Síndrome de La Habana: un caso de guerra de quinta generación

Resumen

Más de mil personas han sido diagnosticadas con el síndrome de La Habana y muchos expertos han llegado a la conclusión de que es causado por un DEW desconocido. El Síndrome de La Habana tiene todas las características de un ataque de Guerra de Quinta Generación (5GW) y la teoría de 5GW puede proporcionar un marco adecuado para analizar la intención de los atacantes, lo que ayudará a contrarrestar y disuadir futuros ataques. 5GW postula que la manipulación de la percepción y las narrativas es fundamental para lograr los objetivos del agresor, que los ataques son encubiertos o ambiguos, que la violencia está muy dispersa y que un objetivo importante es socavar la legitimidad de un Estado y su capacidad para funcionar.

Palabras clave: Síndrome de La Habana, armas de energía dirigida, guerra de quinta generación

doi: 10.18278/gsis.8.1.5

哈瓦那综合征：第五代战争案例

摘要

已有超过一千人被诊断患有哈瓦那综合征，并且许多专家得出结论，认为这是由未知的定向能武器引起的。哈瓦那综合征具有第五代战争(5GW)攻击的所有特征，并且5GW理论能提供合适的分析框架来分析攻击者意图，这将有助于反击和威慑未来袭击。5GW认为，操纵感知和叙事对于实现侵略者的目标而言至关重要，攻击是隐蔽或模糊的，暴力是高度分散的，并且主要目标是破坏国家的合法性及其运作能力。

关键词：哈瓦那综合征，定向能武器，第五代战争

Introduction

Since 2008 there has been a discussion about a potential new Cold War that would again divide the world into East and West engaged in a geopolitical competition with each other.[1] Although the notion of a new Cold War has been disputed during the Obama administration, it is now widely accepted that the West is in a Cold War-like competition with Russia and China, following the Russian invasion of Ukraine and the growing tensions in the Taiwan strait.[2] Already during the Trump administration there were indications of a covert conflict between the West and Russia, as indicated in the Russian interference with the U.S. elections in 2016 and the Skripal affair of 2018.[3]

In late 2016 another phenomenon appeared, namely 'health attacks' on U.S. and Canadian diplomats in Cuba with an unknown mode of attack, which was first reported by the New York Times in August 2017.[4] More US diplomats and intelligence officers around the world became victim of the so-called Havana Syndrome, named after the first location, where this phenomenon has been observed and documented. There have been multiple U.S. government investigations of the phe-

---

1   Edward Lucas, *The New Cold War: Putin's Russia and the Threat to the West* (New York: St. Martin's Press, 2009).

2   Robert Levgold, *Return to Cold War* (Cambridge, UK: Polity, 2016), 89; Hal Brands, *The Twilight Struggle: What the Cold War Teaches Us about Great Power Rivalry Today* (New Haven, CT: Yale University Press, 2022), 2.

3   Rebekah Koffler, *Putin's Playbook: Russia's Secret Plan to Defeat America* (Washington. DC: Regnery Gateway, 2021), XIII, 245-246.

4   Emily Cochrane, "Health Drove US Envoys Out of Cuba," *New York Times*, August 11, 2017.

nomenon coming to different conclusions with speculations ranging from new types of unknown directed energy weapons (DEWs) to crickets and mass hysteria as possible explanations for the Havana Syndrome.[5] Medical research clearly indicates that some individuals have diagnosable and provable physiological symptoms that form a medical syndrome.[6] After many years of observing this phenomenon it is now time to go beyond the question as to whether the Havana Syndrome is real and ask the question as to what objectives may be behind the covert DEW attacks.

Considering the current geopolitical tensions and the Russian track record of conducting covert attacks on U.S. diplomats and intelligence, as well as Russian research in unorthodox weapons technologies, it is a reasonable conclusion that the Havana Syndrome is caused by deliberate covert attacks by Russia or another state or non-state actor, pursuing some strategic objective. This paper argues that the Havana Syndrome fits well into the framework of

Fifth Generation Warfare (5GW) and may be an attempt to sabotage U.S. foreign policy and contribute to the political destabilization of the U.S. The paper will first explain 5GW and interpret it as cognitive warfare, a term suggested by NATO, before discussing the Havana Syndrome. The final step will be to put the covert "health attacks" in the framework of 5GW in order to speculate about the actor(s) and intentions behind using DEW for covert attacks on U.S. personnel.

Fifth Generation Warfare (5GW) as Cognitive Warfare

Since the late 1980s there has been a debate among military professionals and academics as to whether the so-called Clausewitzian concept of war is still adequate or whether war would be fundamentally changing its character to now include conflict short of war, non-political violence, non-violent strategies, cyberwar, and covert modes of conflict.[7] In short, some military theorists sought to broaden

---

5   U.S. Department of State, "Havana Cuba: Accountability Review Board," June 2018, Declassified on October 1, 2019; JASON, "Acoustic Signals and Physiological Effects on U.S. Diplomats on Cuba," MITRE Corporation, November 2018; CDC, "Cuba Unexplained Events Investigation: Final Report," December 3, 2019, available at: <https://nsarchive.gwu.edu/documents/cdc-report-havana-syndrome-medical-mystery-remains-unresolved/CDC%20%2320200983DOS255%20Final%20Report.pdf>; Relman and Pavlin (eds.), *An Assessment of Illness in U.S. Government Employees and Their Families in Overseas Embassies* (Washington, DC: National Academy of Sciences, 2020); "IC Expert Panel on Anomalous Health Incidents Report: Executive Summary," Office of the Director of National Intelligence, declassified February 1, 2022; Robert W. Baloh and Robert E. Batholomew, *Havana Syndrome: Mass Pyschogenic Illness and the Real Story Behind the Embassy Mystery and Hysteria* (Cham, Switzerland: Copernicus, 2020).

6   Randal L. Swanson, Stephen Hampton, Judith Green-McKenzie, Ramon Diaz-Arrastia,Sean Grady, Ragini Verma, Rosette Biester, Diana Duda, Ronald L. Wolf, and Douglas H. Smith, "Neurological Manifestations Among US Government Personnel Reporting Directional Audible and Sensory Phenomena in Havana, Cuba," *JAMA* 319, no. 11 (2019): 1125-1133.

7   For an overview of these ideas see John Arquilla and David Ronfeldt, *In Athena's Camp: Preparing for Conflict in the Information Age* (Santa Monica, CA: RAND, 1997).

the concept of war in order to include measures and modes of conflict that have not traditionally been considered to be war. One of the earliest attempts to redefine war in this manner was the Fourth Generation Warfare school, which postulated that modern warfare can be divided into three distinctive generations characterized by manpower/ mass, firepower/ attrition, and maneuver with a fourth generation emerging that would be based on insurgency or political warfare.[8]

There have been a couple of variations in the Fourth Generation Warfare school, with Martin van Creveld emphasizing the decay of the nation state and the end of the Westphalian order, the New Wars school emphasizing warlordism and the new war economies that keep conflicts going for decades, and the evolved insurgency school that emphasized new aspects in current insurgencies.[9] By the early 2000s some 4GW theorists moved on to declaring the emergence of a fifth generation, which will be shown is focused on dominating the human domain by way of manipulating perceptions.

The Original Understanding of 5GW

5GW was a theoretical response to new modes of terror as observed in the Amerithrax attack and the 9/11 attacks, which indicated that some non-state actors and super-empowered individuals could pose a serious threat to major nations, using commercial technology that has been weaponized.[10] There were a couple of military journal articles in in the 2000s about 5GW, which failed to establish any common definition of 5GW.[11] In 2010 Daniel H. Abbott published a collection of articles in *The Handbook of 5GW*, which brought more clarity and a better theoretical foundation for what 5GW may be.[12] In essence, 5GW is warfare at the societal level that targets a society or civilian sub-groups rather than the military. It is suggested that 5GW would amount to "[m]oral and cultural warfare [that] is fought through manipulating perceptions and altering the context by which the world is perceived."[13] This would be accomplished by clandestine means and a minimum of violence, making it very difficult to recognize 5GW as acts

---

8   William Lind, Keith Nightingale, John Schmitt, Joseph Sutton, and Gary Wilson, 'The Changing Face of War: Into the Fourth Generation', *Marine Corps Gazette*, March 2016, 86-90.

9   Martin van Creveld, *The Transformation of War* (New York: Simon & Schuster, 1991); Mary Kaldor, *New and Old Wars: Organized Violence in a Global Era* (Cambridge, UK: Polity, 2001); and Thomas Hammes, *The Sling and the Stone: On War in the 21ˢᵗ Century* (St. Paul, MN: Zenith Press, 2004).

10   Thomas Hammes, "Fourth Generation Evolves, Fifth Emerges," *Military Review* (May-June 2007): 20.

11   Donald J. Reed, "Beyond the War on Terror: Into the Fifth Generation of War and Conflict," *Studies in Conflict & Terrorism* 31, no. 8 (2008): 689.

12   Daniel H. Abbott (ed.), *The Handbook of 5GW: A Fifth Generation of Warfare?* (Ann Arbor, MI: Nimble Books LLC, 2010).

13   Shane Deichman, "Battling for Perception: Into the 5ᵗʰ Generation?," in: Abbott (ed.), *The Handbook of 5GW*, 12.

of war, which is acknowledged by the authors: "violence is so dispersed that the losing side may never realize that it has been conquered. The very secrecy of 5GW makes it the hardest generation of war to study."[14]

Another key idea in *The Handbook of 5GW* is that belligerents may rely on the manipulation of proxies for attacking their enemies indirectly in order to create confusion about who the real enemy is. According to David Axe, "5GW is when a party exploits or encourages an existing or emerging crisis to achieve strategic goals that those involved in the crisis are not aware of. 5GW is a form of stealthy proxy war."[15] This means that the measures chosen are deliberately ambiguous or covert, so that activities appear to be benign rather than malign, are not visible at all, or are likely to be attributed incorrectly to hide the real aggressor.

Unrestricted Warfare

Another theory closely associated with 5GW originates from two Chinese military officers, who published a book in 1999 with the title *Unrestricted Warfare*.[16] The basic argument in this book is that war is not limited to the employment of military means but would include every and all means that can be used for weakening an adversary with no limitation to what may be done.[17] The authors suggest that "there is nothing in the world today that cannot become a weapon, and this requires that our understanding of weapons must have an awareness that breaks through all boundaries."[18]

Non-military war would include trade wars (attack another country's economy through sanctions), financial wars (stock and currency market manipulation), terrorism (WMD and cyber), ecological war (environmental modification weapons), psychological warfare (spreading rumors to break the enemy's will), smuggling warfare (cause chaos in markets), drug warfare (harm other countries by spreading narcotics addiction), network warfare (cyber warfare), technological warfare (establish technological monopolies), fabrication warfare (complex deceptions), resources warfare (seizing critical resources), economic aid warfare (use aid as leverage over others), cultural warfare (become the leading culture and assimilate other cultures), and international law warfare (use the rules of international law against others).[19]

Both Donald Reed and George Michael have connected *Unrestricted Warfare* to 5GW in the sense that it ex-

---

14   Daniel H. Abbott, 'Introduction', in Abbott (ed.), *The Handbook of 5GW*, 10.

15   David Axe, 'Piracy, Human Security, and 5GW in Somalia', in: Abbott (ed.), *The Handbook of 5GW*, 152.

16   Qiao Liang and Wang Xiansui, *Unrestricted Warfare* (Beijing: PLA Literature and Arts Publishing House, 1999).

17   Liang and Xiansui, *Unrestricted Warfare*, 56, 170.

18   Liang and Xiansui, *Unrestricted Warfare*, 25.

19   Liang and Xiansui, *Unrestricted Warfare*, 50-55.

pands the scope of warfare in terms of the means, the geography, and the types of belligerents (both state and nonstate).[20] The authors of *Unrestricted Warfare* made it clear that criminal means are acceptable in this new type of war, which can be also summed up, as Mark Galeotti suggested, as a "weaponization of everything."[21] Robert Bunker, who reviewed the book compared it to Mao Zedong's *On Guerrilla Warfare* and suggested that "the significance of this work cannot be overstated."[22]

Cognitive Warfare

The term cognitive warfare is the latest descriptor of a new mode of conflict that aims to psychologically manipulate target audiences in support of strategic objectives. It seems to have originated from an article published in 2021 on NATO's website with the title: "Countering Cognitive Warfare: Awareness and Resilience."[23] The authors claim that:

> "[i]n cognitive warfare, the human mind becomes the battlefield. The aim is to change not only what people think, but how they think and act. Waged successfully, it shapes and influences individual and group believes and behaviours to favour an aggressor's tactical or strategic objectives. In its extreme form, it has the potential to fracture and fragment an entire society, so that it no longer has the collective will to resist an adversary's intentions. An opponent could conceivably subdue a society without resorting to outright force or coercion."[24]

Of course, these general ideas are not new and are a staple in what has been traditionally termed psychological operations or Psyops and also what has previously referred to as "hybrid warfare" by NATO or "new generation warfare" by Russian authors.[25] What is new is the idea that cognitive attacks can achieve strategic objectives and may be a substitute for a direct or overt military

---

20   Reed, 'Beyond the War on Terror: Into the Fifth Generation of War and Conflict', 697; George Michael, *Lone Wolf Terror and the Rise of Leaderless Resistance* (Nashville, TN: Vanderbilt University Press, 2012), 157.

21   Mark Galeotti, *The Weaponisation of Everything: A Field Guide to the New Way of War* (New Haven, CT: Yale University Press, 2022).

22   Robert Bunker, "Unrestricted Warfare: Review Essay I," *Small Wars and Insurgencies* 11, no. 1 (2000): 115.9,5 pt

23   Kathy Cao, Sean Glaister, Adriana Pena, Danbi Rhee, William Rong, Alexander Rovalino, Sam Bishop, Rohan Khanna, and Jatin Saini, "Countering Cognitive Warfare: Awareness and Resilience," *NATO Website*, May 20. 2021, available at: <https://www.nato.int/docu/review/articles/2021/05/20/countering-cognitive-warfare-awareness-and-resilience/index.html>.

24   Cao e.a., "Countering Cognitive Warfare."

25   Richards Kols, "NATO Must Meet Russia's Hybrid Warfare Challenge," *Atlantic Council*, July 3, 2018, <https://www.atlanticcouncil.org/blogs/new-atlanticist/nato-must-meet-russia-s-hybrid-warfare-challenge/>; S.G. Chekinov,and S.A. Bogdanov, "The Nature and Content of a New-Generation War," *Military Thought* 10 (2013).

attack, which means that Western liberal societies could "lose the next war without a fight."[26]

There may be also another novel dimension, which is the weaponization of neuroscience and the development of new weapons that target the brain or central nervous system to affect perception, emotion, mental capability, or human performance more generally.[27] Robert McCreight has termed them "NeuroStrike weapons" that could be employed in furtherance of cognitive warfare, which he considers to be "a genuine covert blitzkrieg on the mind and all its associated systems."[28]

> "The basic principle of a suggested **neurostrike weapon** is a fairly simple proposition. It entails a hand held, or platform mounted, mixture of an RF, directed energy pulse or **neurocognitive disrupter, combined with acoustic wave dynamics** which is designed to harm, disable or permanently damage a human brain. It may also adversely affect the brains of several in close proximity to the attack."[29]

McCreight has introduced the concept of the NeuroStrike weapon in the context of the Havana Syndrome, arguing that the US government needs to "devise a system to identify and detect when **NeuroStrike** technologies are being used," so that these incidents can be properly investigated. At this point, the U.S. national security establishment has still to grapple with the difficulty of detecting and verifying attacks, as argued by McCreight:

> "The technology is insidious and consistently defies detection, prevention, medical verification and scientific confirmation aside from episodic reports that an anomaly has occurred and impaired the neurological and cognitive wellbeing of its intended targets. Absent a consensus medical case definition and serious causative technology research, these attacks as reported could easily be discounted as psychotic or delusional events where the complaining individuals were shunted aside as emotionally unstable. We must discern what the exact offending technology is and take steps to reduce and mitigate its continued used in future cases elsewhere."[30]

In addition to the challenge of correctly attributing the attacks, it is very important to also understand the strategic intent behind them, so that appropriate steps can be taken to counter and deter future attacks.

---

26   Fracois du Cluzel, "Cognitive Warfare," *NATO Innovation Hub* (June-November 2020), available at: <https://cognitive-liberty.online/wp-content/uploads/20210122_CW-Final.pdf>.

27   du Cluzel, "Cognitive Warfare," 21.

28   Robert MCreight, "Neuro-Cognitive Warfare: Inflicting Strategic Impact Via Non-Kinetic Effect," *Small Wars Journal*, September 16, 2022.

29   McCreight, "Neuro-Cognitive Warfare."

30   McCreight, "Neuro-Cognitive Warfare."

The Havana Syndrome

Starting from October 2016 several U.S. diplomats in Havana suffered from inexplicable health conditions. Six of them had to be flown from Cuba to Miami for emergency medical treatment. At least 21 American diplomats and some Canadian diplomats suffered symptoms by 2017.[31] Some of the symptoms included headaches, dizziness, insomnia, ear pain, temporary loss of hearing, weakened sense of balance, diminished memory, inability to regulate emotion, and indications of traumatic brain injury and concussion.[32] In December 2017 senior CIA officer Marc Polymeropoulos experienced a sudden onset of nausea in his Moscow hotel room, suffering a second attack two days later in a Moscow restaurant. He had to be flown out of the country and developed permanent symptoms similar to the diplomats in Havana, which forced him to retire from his CIA career of 26 years in 2019.[33]

Several diplomats in Guangzhou, China suffered from similar symptoms, including brain injury in spring 2018.[34] It had started with one embassy employee hearing sounds and getting sick in late 2017 with more cases appearing in April 2018. Mark Lenzi, a security engineer at the consulate had to be flown out of China together with his family.[35] Further incidents involving U.S. and Canadian government personnel (diplomats, intelligence officers, and staffers) were reported in several locations around the world, including Cuba, China, Russia, Austria, India, Vietnam, Colombia, Uzbekistan, Kirgizstan, and Washington, DC.[36] Potential attacks occurred in a variety of settings such as "residencies, on the street, in vehicles, and even at U.S. secure facilities."[37] The incidents have seemingly escalated in geographic scope, frequency, and severity since late

---

31 A. Erickson, "All the Theories What's Happening to the Diplomats in Cuba: At Least 21 Americans Have Reported a Bizarre Rash of Symptoms," *Washington Post Blogs*, September 30, 2017, available at: <https://www.washingtonpost.com/news/worldviews/wp/2017/09/30/all-the-theories-about-whats-happening-to-the-diplomats-in-cuba/>.

32 F. Robles and K. Semple, 'U.S. and Cuba Baffled by "Health Attacks" on American Envoys in Havana,' *New York Times*, August 12, 2017, A6.

33 J. Ioffe, "The Mystery of the Immaculate Concussion," *GQ Magazine*, October 20, 2020, available at: <https://www.gq.com/story/cia-investigation-and-russian-microwave-attacks>.

34 H. Gardiner, "25th Person at U.S. Embassy Is Mysteriously Sickened," *New York Times*, June 21, 2018, available at: <https://www.nytimes.com/2018/06/21/us/politics/us-diplomat-cuba-embassy-illness.html>.

35 S.L. Meyers and J. Perlez, "U.S. Diplomats Evacuated in China as Medical Mystery Grows," *New York Times*, June 6, 2018.

36 Jack Dutton, "Havana Syndrome Symptoms Have Been Reported in these Countries," *Newsweek*, September 21, 2021, available at: <https://www.newsweek.com/havana-syndrome-china-austria-russia-1631102>.

37   Paul Kolbe, Marc Polymeropoulus, and John Sipher, "Havana and the Global Hunt for U.S. Officers," *The Cipher Brief*, October 24, 2021, available at: <https://www.thecipherbrief.com/havana-syndrome>.

2016.[38] Altogether over 1,000 people have been diagnosed with Havana Syndrome as of February 2022.[39]

Mysterious "Anomalous Health Attacks"

The State Department conducted an investigation and arranged for the medical examination of the affected diplomats to determine the cause of the symptoms. The University of Pennsylvania Center for Brain Injury and Repair found a number of conditions that could be objectively tested, such as evidence of concussion in persons with no history of brain injury and temporary auditory impairment.[40] With the help of neuroimaging it was found that there were significant "differences in whole brain white matter volume, regional gray and white matter volume, cerebellar tissue microstructural integrity, and functional connectivity in the auditory and visuospatial subnetworks but not in the executive control subnetwork" compared to a control group.[41]

The existence of objective physiological conditions discovered in the majority of the affected diplomats should rule out some kind of mass psychosis as the cause, although this claim is implausibly still being made by some researchers.[42] Secretary of State Rex Tillerson characterized the incident as health attacks and there was initial speculation that the cause was a new type of sonic or acoustic weapon that had been developed by Russia or China.[43] An alternative explanation was that there was no intent to harm the diplomats but that a new type of eavesdropping equipment installed by the Cubans in the U.S. embassy had caused the symptoms.[44] The Trump administration took the matter seriously enough to expel 15 Cuban diplomats in retaliation for the

---

38 U.S. Congress, H.R.4914 passed on August 3, 2021, available at: <https://www.congress.gov/117/bills/hr4914/BILLS-117hr4914ih.pdf>, Section 2 (13).

39 Ken Dilanian, "'Havana Syndrome' Symptoms in Small Group Most Likely Caused by Directed Energy, Says U.S. Intel Panel of Experts,' *NBC News*, February 2, 2022, available at: <https://www.nbcnews.com/politics/national-security/havana-syndrome-symptoms-small-group-likely-caused-directed-energy-say-rcna14584>.

40 Swanson e.a., "Neurological Manifestations Among US Government Personnel Reporting Directional Audible and Sensory Phenomena in Havana."

41 R. Verma, R.L. Swanson, D. Parker, A.A. Ould Ismail, R.T. Shinohara, J.A. Alapatt, J. Doshi, "Neuroimaging Findings in U.S. Government Personnel with Possible Exposure to Directional Phenomena in Havana, Cuba," *Journal of the American Medical Association* 322, no. 4 (2019): 336-347. doi:10.1001/jama.2019.9269.

42 R.W. Baloh and R.E. Bartholomew, *Havana Syndrome: Mass Psychogenic Illness and the Real Story Behind the Embassy Mystery and Hysteria* (Cham, Switzerland: Copernicus, 2020).

43 Robles and Semple, 'U.S. and Cuba Baffled by "Health Attacks" on American Envoys in Havana.'

44 T. Wilkinson, "Did Covert Spy Gear Cause U.S. and Canadian Diplomats in Cuba to Lose Their Hearing?," *Los Angeles Times*, August 10, 2017, available at: <https://www.latimes.com/nation/la-fg-cuba-hearing-20170810-story.html>.

health attacks.[45]

Some of the affected diplomats have sought legal remedies to get compensation from the government. Mr. Zaid, who is a lawyer representing some of them, stated: "'It's sort of naïve to think this just started now"... Globally, he added, covert strikes with the potent beams appear to have been going on for decades'.[46] Furthermore, the National Security Agency gave Zaid 'a statement on how a foreign power built a weapon "designed to bathe a target's living quarters in microwaves, causing numerous physical effects, including a damaged nervous system.'[47]

Congress recently awarded long-term emergency health benefits to the affected diplomats, thereby acknowledging that their ordeal was serious, significant, real, and not imagined.[48] Affected diplomats have been awarded a lump sum compensation between $140,000 and $187,000 for hardship, permanent injury received on duty, and subsequent loss of employment.[49] More details are contained in House of Representatives passed Bill H.R.4914 on August 3, 2021 that authorized the compensation. The bill states as rationale: "To impose sanctions against foreign persons and foreign governments in response to certain clandestine attacks on United States personnel, and for other purposes."[50] In particular it suggested, "United States personnel have suffered persistent brain injuries after being targeted in attacks that have been increasing in number, geographic location, and audacity."[51] The covert attacks would be "continuing and [the] expanding scope of these attacks has become a serious security concern that is also undermining the morale of United States personnel, especially those posted at overseas diplomatic missions."[52]

The Investigations

The U.S. State Department tasked the National Academy of Sciences (NAS) to advise on how U.S. diplomats could be protected. A standing committee with 19 scientists led by David Relman, a professor of microbiology at Stanford University, reviewed and studied

45   Meyers and Perlez, "U.S. Diplomats Evacuated in China as Medical Mystery Grows."

46   W. Broad, "Microwave Weapons Are Prime Suspect in Ills of U.S. Embassy Workers," *New York Times*, September 1, 2018.

47   Broad, "Microwave Weapons Are Prime Suspect in Ills of U.S. Embassy Workers."

48   Reuters, "'Havana syndrome" U.S. Diplomats Get Benefits in Spending Bill,' *Reuters*, December 16, 2019, available at: <https://www.reuters.com/article/us-usa-budget-congress-diplomats/havana-syndrome-u-s-diplomats-get-benefits-in-spending-bill-idUSKBN1YK24Z>.

49   Kylie Atwood, Katie Bo Lillis, and Jennifer Hansler, 'Biden Administration to Compensate Some "Havana Syndrome" Victims Up to $187,000,' *CNN*, June 24, 2022, available at: <https://www.cnn.com/2022/06/23/politics/havana-syndrome-victims-compensation/index.html>.

50   U.S. Congress, H.R.4914 passed on August 3, 2021, available at: https://www.congress.gov/117/bills/hr4914/BILLS-117hr4914ih.pdf.

51   U.S. Congress, H.R.4914 passed on August 3, 2021, Section 2 (1).

52   U.S. Congress, H.R.4914 passed on August 3, 2021, Section 2 (13).

the Havana syndrome and it produced a final report that was declassified in December 2020.[53] Although not explicitly asked to resolve the mystery of the Havana syndrome, the committee's charter still included the investigation of plausible mechanisms that could have caused the reported symptoms. The committee looked at directed radio frequency energy, chemicals, infectious agents, and psychological and social factors.[54] The committee had access to the clinical diagnoses of all victims by the NIH and interviewed eight of them. Sound as a mechanism was apparently ruled out early on by the committee as there is no discussion of sonic weapons in the report. Chemicals were ruled out because the observed symptoms were not consistent with exposure to some kind of nerve agent or pesticide.[55] Biological agents were ruled out on similar grounds. The committee considered delusions and hallucinations that may have been caused by stress but concluded that "[t]he acute initial, sudden-onset, distinct and unusual symptoms and signs described in some affected DOS

personnel (see Section 3 and CDC Report) cannot be ascribed to psychological and social factors in the absence of patient-level data."[56]

They suggested that "many of the cognitive, vestibular, and auditory effects observed in DOS personnel are most consistent with modulated, or pulsed, RF biological effects," citing Soviet/ Russian research on pulsed and continuous RF energy on humans.[57] Notably, the report described a lack of comparable U.S. research regarding RF energy on humans and the NAS committee did not have access to classified research.[58] The report points at known RF effects such as the

"perceived clicking sound within the head even when the ears were covered, a perceived force/pressure sensation within the head and on the face, perceived spatial localization and directionality of these perceived phenomena and other loud sounds, hearing loss, tinnitus, impaired gait and loss of balance, as well as the absence of heating sensation and absence

---

53   National Academy of Sciences, "New Report Assesses Illnesses Among U.S. Government Personnel and Their Families at Overseas Embassies," *Engineering Medicine News Release*, December 5, 2020, available at: <https://www.nationalacademies.org/news/2020/12/new-report-assesses-illnesses-among-us-government-personnel-and-their-families-at-overseas-embassies>.

54   Relman and Pavlin (eds.), *An Assessment of Illness in U.S. Government Employees and Their Families in Overseas Embassies.*

55   Relman and Pavlin (eds.), *An Assessment of Illness in U.S. Government Employees and Their Families in Overseas Embassies*, 23.

56   Relman and Pavlin (eds.), *An Assessment of Illness in U.S. Government Employees and Their Families in Overseas Embassies*, 28.

57   Relman and Pavlin (eds.), *An Assessment of Illness in U.S. Government Employees and Their Families in Overseas Embassies*, 18.

58   Relman and Pavlin (eds.), *An Assessment of Illness in U.S. Government Employees and Their Families in Overseas Embassies*, X.

of observed disruption of electronic devices in the immediate environment."[59]

The microwave hearing or Frey-effect mentioned in the report was first discovered in the 1940s and has been documented in scientific research since 1962.[60] Pulsed RF frequencies above 5kHz can stimulate the cochlea similar to auditory stimuli and can cause buzzing, clicking, and chirping sounds, as well as tinnitus, all of which was reported by Havana Syndrome victims.[61] Referring to other studies in the field the report states: "Pulsed RF effects on the nervous system can include changes to cognitive..., behavioral..., vestibular..., EEG during sleep..., and auditory... function in animals and humans."[62]

The findings of the NAS report are also in line with older research conducted on nonlethal technologies by the Department of Defense from the 1990s, which described the Frey effect and other thermal microwave effects, such as disorientation, disruption of memory, hearing loss, and brain damage.[63]

Thus, the NAS report goes beyond the older research in the sense that it is one of the rare times when the bioeffects of RF energy are scientifically acknowledged in western mainstream science. The findings are in line with the growing recognition in the Western scientific community that electrosensitivity, which has the widely acknowledged symptoms of headaches, tinnitus, dizziness, tinnitus, and cognitive defects, is a physiological and not a psychosomatic condition. This means that RF exposure at levels that are officially considered safe can harm some people, which can be also objectively tested.[64]

The JASONs, the FBI, the CDC, CIA, and ODNI have conducted separate investigations and their conclusions vary. The report by the DOD scientific advisory group JASON was published relatively early in 2019 essentially dismissed RF and microwaves as a source and instead suggested the recorded sounds matched those of crickets.[65] The FBI was sent to Cuba to investigate the cases after the affected diplomats had left and had the FBI Behavioral Analysis

59   Relman and Pavlin (eds.), *An Assessment of Illness in U.S. Government Employees and Their Families in Overseas Embassies*, 18.

60   A.H. Frey, "Human Auditory System Response to Modulated Electromagnetic Frequency," *Journal of Applied Physiology* 17, no. 4 (1962): 689-692.

61   A. Elder and C.K. Chou, "Auditory Response to Pulsed Radiofrequency Energy," *Bioelectromgnetics Supplement* 6 (2003): S.162-S.173. doi:10.1002/bem.10163.

62   Relman and Pavlin (eds.), *An Assessment of Illness in U.S. Government Employees and Their Families in Overseas Embassies*, 18.

63   U.S. Department of Defense, "Bioeffects of Selected Nonlethal Weapons," *Department of the Army* (1998), declassified on December 6, 2006.

64   D. Belpomme and P. Irigaray, "Electrohypersensitivity as a Newly Identified and Characterized Neurologic and Pathologic Disorder: How to Diagnose, Treat, and Prevent It," *International Journal of Molecular Sciences* 21, no. 1915 (2020): 1-20. doi:10.3390/ijms21061915.

65   JASON, "Acoustic Signals and Physiological Effects on U.S. Diplomats on Cuba."

Unit conduct an assessment of the victims. They concluded that they suffered from mass psychogenic illness, but the FBI had not actually spoken to any of them, nor did they consider the physiological symptoms.[66] The FBI has since acknowledged that some of their agents have symptoms consistent with the Havana Syndrome.[67] The CDC released a report on the Havana Syndrome in late 2019, which merely reviews the research undertaken by the University of Miami, the University of Pennsylvania, and NIH regarding 95 documented cases. It concluded that "the evaluations conducted thus far have not identified a mechanism of injury."[68]

An interim report by the CIA suggested that foreign governments are unlikely to be responsible for the Havana Syndrome. According to *Politico*, the report stated 'that the vast majority of reported cases can be explained by medical, environmental or technical factors—including previously undiagnosed illnesses—and that it is "unlikely" that a malicious state actor is inflicting

purposeful harm on U.S. diplomats on a far-reaching, worldwide scale.'[69] The CIA seems to have backpaddled a bit from its initial conclusions. CIA Director Burns suggested that the Havana Syndrome was "real," that the role of foreign actor has not been ruled out, and that the investigation was ongoing.[70]

Finally, a report produced by an ODNI IC Experts Panel on Anomalous Health Incidents claimed in the declassified Executive Summary that "[t]he signs and symptoms of AHIs are genuine and compelling" that "cannot be easily explained by known environmental or medical conditions," and that "[p]ulsed electromagnetic energy, particularly in the radiofrequency range, plausibly explains the core characteristics."[71] James Giordano, an expert in the field of military neuroscience, similarly suggested that

> "[t]he most likely culprit here would be some form of electromagnetic pulse generation and/or hypersonic generation that would then utilize the ar-

66   Adam Entous, "Stealth Mode: How the Havana Syndrome Spread to the White House," *The New Yorker* 97, no. 14 (May 24, 2021).

67   Ken Dilanian, "FBI Acknowledges Some Agents May Have Havana Syndrome Symptoms," *NBC News*, November 24, 2021, available at: <https://www.nbcnews.com/health/health-news/fbi-acknowledges-agents-may-havana-syndrome-symptoms-rcna6504>.

68   CDC, "Cuba Unexplained Events Investigation," 14.

69   Alexander Ward and Andrew Desiderio, 'U.S. Foe or Specific Weapon Not Behind Sustained, Global Campaign Causing "Havana Syndrome," CIA Finds,' *Politico*, January 19, 2022, available at: <https://www.politico.com/news/2022/01/19/havana-syndrome-cia-causes-527457>.

70   Julian Barnes, "C.I.A. Officer Suffers Havana Symptoms," *New York Times*, September 20. 2021; Conor Finnegan and Cindy Smith, "CIA Says Foreign Actor May Be Behind Some Havana Syndrome Cases," *NBC News*, January 20, 2022, available at: <https://abcnews.go.com/Politics/cia-foreign-actor-havana-syndrome-cases/story?id=82376545>.

71   U.S. Office of the Director of National Intelligence, "IC Expert Panel on Anomalous Health Incidents Report: Executive Summary," declassified February 1, 2022, 2.

chitecture of the skull to create something of an energetic amplifier or lens to induce a cavitational effect that would then induce the type of pathologic changes that would then induce the constellation of signs and symptoms that we're seeing in these patients."[72]

Brain scans of several of the victims have indicated concussion, as well as other symptoms that are consistent with a DEW attack.[73] Another rationale as to why some analysts and commentators believe that the Havana Syndrome may be caused by a deliberate attack is that it would not be the first time U.S. diplomats and intelligence officers have been subjected to covert electromagnetic exposure.[74]

Echoes from the Past

U.S. diplomats were subjected to microwave bombardment from the 1950s to the mid-1970s. The CIA conducted routine bug sweeps in the new U.S. embassy and detected the so-called Moscow-Signal as early as 1953. A single microwave beam from an apartment building on the opposite side of the street radiated the embassy with no clear regularity. By the late 1950s the signal became continuous.[75] By 1962 it was discovered that the microwaves were targeting specifically the offices of the ambassador and top intelligence officials.[76] The frequencies used by the Soviets varied and were in the range of 2 GHz and 7 GHz with a maximum intensity of 0.005 milliwatts before May 1975.[77] The State Department kept the matter secret from the embassy employees but made a series of diplomatic protests over the microwave exposure, starting from 1967.[78]

The Moscow signal was first publicly reported by Jack Anderson in 1972 with the speculation by the CIA that the Soviets might try to "brainwash" U.S. diplomats.[79] The microwave radiation increased dramatically in intensity from 1973, which was said to have resulted in direct physical harm and the psy-

---

72 Steve Dorsey, 'Pentagon Turns Focus to Cuba Health "Attacks" Amid New Findings on American Victims,' *CBS News*, September 12, 2018, available at: <https://www.cbsnews.com/news/pentagon-turns-focus-to-cuba-attacks/>.

73 Rita Rubin, "More Questions Raised by Concussion-like Symptoms Found in US Diplomats Who Served in Havana." *JAMA* 319, no. 11 (2018): 1080; Julian Barnes, "Report Criticizes C.I.A. Handling of Havana Syndrome Cases," *New York Times*, October 14, 2022.

74 Jamey Essex and Joshua Bowman, "From the Green Zone to the Havana Syndrome: Making Geographic Sense of Rotationality and Hardship in Diplomacy," *Diplomatica* 4, no. 1 (2022): 93.

75 H. Pollack, "Epidemiologic Data on American Personnel in the Moscow Embassy," *Symposium on Health Aspects of Non-Ionizing Radiation*, New York, April 9-10, 1979.

76 A. Jacobsen, *Phenomena: The Secret History of the U.S. Government's Investigation Into Extrasensory Perception and Psychokinesis* (New York: Little Brown and Company, 2017), 74.

77 Pollack, "Epidemiologic Data on American Personnel in the Moscow Embassy," 1183.

78 B. Reppert, "Zapping an Embassy: 35 Years Later, the Mystery Lingers," *Times Daily*, May 22, 1988, 6D.

79 J. Anderson, "'Brainwash' attempt by Russians?,' *Washington Post*, May 10, 1972, B15.

chological trauma of U.S. embassy personnel.[80] In some areas of the embassy the EMF exposure had increased from an average 0.001 milliwatt (mW) per square centimeter to up to 0.018 mW/cm², prompting the State Department to block the microwaves with aluminum screens in early 1976.[81] Reportedly, many U.S. diplomats in the Moscow embassy suffered from ill health and several died from cancer, including two U.S. ambassadors. Ambassador Stoessel eventually threatened to resign over the concerns of his staff, and he eventually died ten years later of leukemia in 1986.[82] However, the RF exposure was well within the levels of what was considered safe, both in the U.S. and the Soviet Union—the U.S. standard is 10 mW/cm², adding to the mystery.[83]

Johns Hopkins University completed a study on the Moscow signal in 1978, known as the Lilienfeld study and found no evidence of a higher cancer rate among U.S. personnel. A study by John R. Goldsmith that reviewed the Lilienfeld study suggested that there were problems with the methodology and that the study may have been biased from the start due to the U.S. Department of State funding it and trying to avoid ethical and legal responsibility for keeping the microwave exposure secret from embassy personnel.[84]

At the time, there were three major theories as to why the Soviets radiated the U.S. embassy with microwaves: 1) they tried to disrupt U.S. electronic surveillance equipment and activities from the embassy; 2) they tried to eavesdrop on the U.S. embassy; and 3) they tried to psychologically, behaviorally, or physiologically affect the U.S. diplomats.[85] The Moscow Signal resulted in Pentagon-funded research into biological and psychological effects of RF, codenamed Project Pandora, from 1965 to 1970.[86] The eventual findings of Project Pandora were inconclusive.[87] The true motivations of the Soviets have remained a mystery, although they themselves admitted to the microwave irradiation of the embassy and suggested that it was for the purpose of interfering with electronic equipment at the embassy.[88] Russian military authors

80   R. Evans and R. Novak, "The Microwave Affair," *Washington Post*, March 8, 1976, A19.

81   New York Times, "Embassy Radiation Is Cut in Moscow," *New York Times*, April 26, 1976, 5.

82   J.C. Lin, "The Moscow Embassy Microwave Signal," *Radio Science Bulletin* 363 (2017): 90-93.

83   J. Martinez, 'The "Moscow Signal" Epidemiological Study, 40 Years On,' *Reviews on Environmental Health* 34, no.1 (2019): 13-24. doi:10.1515/reveh-2018-0061, 14.

84   J.R. Goldsmith, "Balancing the Interests of Patients, Science and Employees: Case Study of RF (Microwave) Exposure of US Embassy Staff in Eastern European Posts," *The Science of the Total Environment* 184 (1996): 87.

85   Reppert, "Zapping an Embassy: 35 Years Later, the Mystery Lingers."

86   Jacobsen, *Phenomena*, 75.

87   R. Beeston, "Russian Micro-waves Plan "to Drive U.S. Envoys Mad," *London Daily Telegraph*, May 11, 1972; Anderson, '"Brainwash" attempt by Russians?'

88   James Schumaker, "Before Havana Syndrome, There Was Moscow Signal," *Foreign Service Journal* (January/ February 2022), available at: <https://afsa.org/havana-syndrome-there-was-moscow-sig-

have suggested that Russia/ the Soviet Union had developed electromagnetic weapons that interfered with the brain and central nervous system, a class of weapons that had been previously referred to as "psychotronic weapons."[89]

The Potential Objectives Behind the "Health Attacks"

Assuming that the Havana Syndrome is caused by deliberate attacks using a weapon or mechanism that remains unknown to the public and the Western scientific community, the question then becomes: who is responsible and what is the intent behind the attacks? The 5GW or cognitive warfare framework may be able to provide some direction. A secret weapon that has effects not immediately observable and that are ambiguous enough to be blamed on other causes is perfect for attacks that are covert and almost completely deniable. As argued by Abbott, "[a] fifth Generation War might be fought with one side not knowing who it is fighting. Or even, a brilliantly executed 5GW might involve one side being completely ignorant that there ever was a war."[90] David Ignatius also observed that "[t]hese mysterious attacks are a policymaker's nightmare. You can't accuse another country of warlike assaults without solid facts; the Iraqi WMD fiasco taught a generation of intelligence analysts that lesson. But if you don't hold rogue actors accountable, how do you deter future attacks?"[91] If Russia or another actor had NeuroStrike weapons at their disposal, then they could be used for achieving a number of different objectives.

Deniable Attacks as Retribution

The covert attacks on US government personnel overseas, and in a few cases domestic, could amount to what Martin Libicki has described as non-obvious warfare, which seems to overlap with the concept of 5GW. The purpose of covert attacks may be signaling to the other side: "if you do this bad things will happen to you."[92] The difficulty in this approach is obviously that an ambiguous signal could be misunderstood by the recipient and/or could be falsely attributed to an innocent party. At this point it is mere speculation, but Russia may have both a motive and the capability for such kind of a covert attack.[93] The timing and locations of the attacks may provide more clues.

---

nal>.

89   Timothy Thomas, "The Mind Has No Firewall," *Parameters* (Spring 1998): 84-92.

90   Daniel H. Abbott, "Go Deep: OODA and the Rainbow of xGW," in Abbott (ed.), *The Handbook of 5GW*, 180.

91   David Ignatius, 'Dealing with "Havana Syndrome" Is a Policymaker's Nightmare,' *Washington Post*, October 28. 2021.

92   Martin Libicki, "The Specter of Non-Obvious Warfare," *Strategic Studies Quarterly* (Fall 2012): 96.

93   Sean Power and Michael Miner, "Report – Havana Syndrome: American Officials Under Attack," *Belfer Center Harvard University*, November 4, 2021, available at: <https://www.belfercenter.org/publication/report-havana-syndrome-american-officials-under-attack>.

The U.S. diplomatic relationship has soured since 2012 when Putin returned to the presidency. The Russians have complained that the "color revolutions" in Eastern Europe and in the Middle East were instigated by Western governments and their proxies in order to advance U.S. geopolitical objectives. In particular, the Russian government has accused the West of seeking to destabilize Russia in order to install a pro-Western regime in Moscow and potentially divide Russia so that it becomes geopolitically insignificant.[94] The Russian government believes that it is under an informational attack by the West.[95] Hence, the Russian government may consider it a just response by attacking individuals associated with the U.S. government and key U.S. allies, in particular diplomats and intelligence personnel.

The attacks seemingly started in Cuba at a time when the diplomatic relationship between the U.S. and Cuba started to improve under the Obama administration.[96] Given the long-lasting ties of Cuba and Russia, which endured even after the collapse of the Soviet Union, it is likely that Cuba would be a suitable location for Russian intelligence to target U.S. diplomats as the Cuban government would not interfere with such actions. Conducting attacks against U.S. diplomats outside of Russia would provide the Russian government with plausible deniability. At the same time, conducting attacks in the country of a Russian partner can still send the intended message, namely that Russia disapproves of U.S. diplomacy and seeks to punish the U.S. by attacking U.S. personnel involved in the planning and implementation of U.S. foreign policy.

Disruption of US Foreign Policy

A second reason for the covert attacks could be that the Russian government seeks to disrupt U.S. foreign policy. First of all, diplomats and intelligence officers who get cognitively impaired are no longer able to do their job well. They have to leave the countries they have been posted to, which can disrupt the conduct of U.S. foreign policy. The more U.S. government personnel is harmed, the more difficult it becomes to find qualified personnel to fill positions in certain locations, especially if it becomes widely known that there is high risk of harmful exposure of some kind. The U.S. Department of State had to offer diplomats the highest rate for hardship and cut stuff from 54 to 18 at the U.S. embassy in Cuba and the CIA shut down their station in Havana completely in 2017.[97] The Havana Syndrome allegedly "'dramatically hurt' morale in the diplomatic service and affected

---

94   Nicolas Bouchet, 'Russia's "Militarization" of Colour Revolutions,' *Policy Perspectives* 4, no. 2 (2016): 2-3.

95   Stephen Blank, "Russian Information Warfare as Domestic Counterinsurgency," *The Journal of the National Committee on American Foreign Policy* 35, no. 1 (2013): 31-34.

96   Essex and Bowman, "From the Green Zone to the Havana Syndrome,' 94.

97   Essex and Bowman, "From the Green Zone to the Havana Syndrome," 94; Entous, "Stealth Mode."

recruitment' for the foreign service.[98] Over a thousand U.S. personnel has been affected with at least some suffering permanent injury or disability.[99]

Particularly concerning are attacks that seem to target U.S. personnel in the proximity of senior officials or at locations close to the offices of the U.S. political leadership such as in Washington, DC. For example, an unnamed senior NSC staffer in the Trump administration claims to have been covertly targeted with a DEW near the White House in an attack that lasted several minutes and caused him to believe that he would die.[100] Another incident involved a CIA officer, who was travelling with CIA Director William Burns to India when he suffered a mysterious health condition consistent with the Havana Syndrome.[101] A diplomat visit of Vice President Kamala Harris to Vietnam was delayed for several hours due to the occurrence of a potential covert attack on two U.S. personnel at the Hanoi embassy, who had to be medevacked.[102]

It is fair to say that the Havana Syndrome has already impacted U.S. diplomacy due to disruptions caused by diplomats who have to be sent home because of unexplained illnesses they suffer in the country, the impact on morale and recruitment for the foreign service, and due to the perceived threat to senior U.S. officials, whose visits to certain foreign countries may put them at a higher risk.

Deception

The attacks could be conducted in a manner as to implicate an innocent third party in order to damage U.S. diplomatic relations with a particular state or even cause a conflict between two states. Most commentary that suggests that the Havana Syndrome is caused by deliberate attack with an unknown type of DEW point towards Russia as the most likely culprit.[103] What if the Russian government was innocent and some other party is responsible? The concept of so-called false flags has received some attention in relation to both the conflicts in Syria and Ukraine with both sides alleging that the other side carried out attacks that they then blamed on the enemy. It has been also claimed that "[t]he Kremlin's comfort with trafficking fabricated intelligence

98   Julian Borger, 'Havana Syndrome has "Dramatically Hurt" Morale, US Diplomats Say,' *The Guardian*, February 10, 2022, available at: <https://www.theguardian.com/us-news/2022/feb/10/havana-syndrome-cuba-us-diplomats-afsa>.

99   Olivia Gazis, 'State Department, CIA Establish Federal Payment Rules for "Havana Syndrome" Victims,' *CBS News*, June 24, 2022, available at: <https://www.cbsnews.com/news/havana-syndrome-state-department-cia-establish-federal-payment-rules-for-victims/>.

100 Entous, "Stealth Mode."

101 Barnes, "C.I.A. Officer Suffer Havana Syndrome Symptoms."

102 Michele Kelemen and Deepa Shivaram, 'VP Harris' Flight Delayed After Possible "Havana Syndrome" Incident in Hanoi,' *NPR*, August 24, 2021, available at: <https://www.npr.org/2021/08/24/1030663913/kamala-harris-havana-syndrome-vietnam-delay-embassy>.

103 Kolbe, Polymeropoulus, and Sipher, "Havana and the Global Hunt for U.S. Officers."

and using false flags as pretexts for coercive acts reflects a well-established track record."[104] This suggests that false flags are used by major powers and can shape narratives and perceptions.

Libicki explored the strategic utility of false-flags: "if an attacker can persuade the target that it was hit by a third party, it may catalyze conflict that will be to the attacker's advantage. A non-obvious Taiwanese cyber attack on the United States during a crisis with China, for instance, might put the United States at odds with China and thus more likely to support Taiwan."[105] Libicki also suggested that covert and unattributed attacks can be carried out for narrative effect: "if the attacker is unknown, or at least unclear, then the focus of the story is necessarily on why the target was attacked—and may well dwell on what the target did to deserve the attack or why the target could not secure itself."[106]

Perhaps a third country is trying to implicate Russia as the culprit for these covert attacks in order to instigate a conflict between the U.S. and Russia that could benefit the true aggressor. If the Havana Syndrome are a false-flag of some kind, the likely perpetrator would be China for a number of reasons: 1)

China likely has the technology for a NeuroStrike weapon as its research activity in the field of high-powered microwaves would dwarf both U.S. and Russian research;[107] 2) a war or intensified competition between the U.S. and Russia could work in China's favor as it would weaken both the U.S. and Russia; and 3) the Chinese government seems to have embraced an unrestricted warfare approach that includes illegal and covert means for attacking the U.S.

It is important not to overlook non-state actors as potential perpetrators. According to 5GW theory non-state actors could be empowered by new technology or could be able to utilize state-like capabilities that they either stole from a state actor or received from a state actor in order to function as their proxy for reasons of deniability. As Hammes pointed out: "The anthrax attacks provided stark evidence that today a single individual can attack a nation state."[108] DEW are not in principle out of reach for non-state actors. In fact, non-state actors have already used simple DEWs such as blinding lasers, which was pointed out in the recent "Directed Futures 2060" study of the Department of Defense.[109] This means, that although a DEW that can

---

104 Huw Dylan and Thomas J. Maguire, "Secret Intelligence and Public Diplomacy in the Ukraine War," *Survival* 64, no. 4 (2022): 37.

105 Libicki, "The Spector of Non-Obvious Warfare," 98.

106 Libicki, "The Spector of Non-Obvious Warfare," 97-98.

107 Edl Schamiloglu, "Experts Believe US Embassies Were Hit With High-Power Microwaves – Here's How the Weapons Work," *SciTechDaily*, February 12, 2022, available at: <https://scitechdaily.com/experts-believe-us-embassies-were-hit-with-high-power-microwaves-heres-how-the-weapons-work/>.

108 Hammes, 'Fourth Generation Evolves, Fifth Emerges', 21.

109 U.S. Department of Defense, 'Directed Energy Futures 2060: Visions for the Next 40 Years

covertly injure personnel in a manner that is hard to detect and protect against would be very difficult to develop, it is not outside of the realm of possibility that a non-state actor could have somehow acquired this capability and use it for the pursuit of their own objectives. A conceivable objective could be that a non-state actor wants to make the U.S. government look weak and helpless in the face of an unknown and unattributable threat, which could delegitimize the U.S. government in the eyes of the world when seemingly unfounded allegations are made against other governments and contribute to general political paranoia and instability in the U.S. As McIntosh argued, "[i]f a 5GW is successful, a target state will have so lost its legitimacy that it cannot be certain of anyone's primary loyalty."[110]

Conclusion

Previous attempts of explaining the Havana Syndrome away as noise produced by crickets, mass psychogenic illness, or health issues with no external cause, have become increasingly unconvincing and should be discarded. The great majority of studies and informed commenters make it clear that the Havana Syndrome must be investigated further in order to properly understand the causes and potentially attribute the attacks in order to deter

future attacks. If the Havana Syndrome is caused by a deliberate attack against U.S. and Canadian personnel, the question becomes what is the intent behind the attacks and what would be a suitable framework for understanding the strategic objectives of the aggressor?

It has been argued in this paper that covert attacks with an unknown DEW can be understood as 5GW, which is a mode of conflict short of open war that relies on secrecy, deception, and influence "in order to make the enemy do our will," using a minimum amount of force.[111] In 5GW, "[t]he conflict is not to conquer the state, or divide the state, but to undermine the state."[112] The covert attacks leverage secret NeuroStrike weapons that utilize Nano-Bio-Info-Cogno (NBIC) technologies for attacking the brain or mental capacity of some U.S. diplomatic and intelligence personnel. The motivation could be revenge or ambiguous signaling, an attempt to disrupt U.S. foreign policy, or an attempt to implicate a third party in order to cause the U.S. to initiate a war or intense competition with that third party or to fuel paranoia and make the U.S. government appear helpless or reckless.

5GW is a war of information, narratives, and perception. In its extreme form it may target an adversary

---

ofU.S.DepartmentofDefenseDirectedEnergyTechnologies', (2021), available at: <https://www.afrl.af.mil/Portals/90/Documents/RD/Directed_Energy_Futures_2060_Final29June21_with_clearance_number.pdf?ver=EZ4QY5MG5UK2LDdwiuPc6Q%3D%3D>, 20.

110 Daniel McIntosh, "Transhuman Politics and Fifth Generation War," in: Daniel H. Abbott, *The Handbook of 5GW*, 82.

111 Daniel H. Abbott, "The xGW Framework," in: Daniel H. Abbott, *The Handbook of 5GW*, 21.

112 Daniel McIntosh, "Transhuman Politics and Fifth Generation War," 81.

brain and mental capacity to more directly impact decision-making in a manner that benefits the aggressor. The uncertainty surrounding the Havana Syndrome breeds paranoia across the U.S. government and may cause paralysis as policymakers are unable to properly comprehend and respond to the threat. As long as the phenomenon remains unexplained, the aggressor enjoys a clear advantage since accusations can be easily denied and since the attacked government is unable to take any overt punitive action in response to the covert attacks since they would appear to be aggressor in the absence of evidence of a prior attack.

The only way to change this situation is to provide incontrovertible scientific evidence for a covert DEW attack and to make this evidence public. Any aggressor engaging in these covert attacks would then be forced to either stop them or face the consequences. Unfortunately, any government that possessed this technology would be unwilling to publicly disclose its existence, as this would severely limit the military usefulness of the technology. The U.S. may already have weapons that can cause the effects associated with Havana Syndrome and may have already secretly deployed these weapons, according to investigative journalist Nicky Woolf.[113] As long as this dynamic is in play it is unlikely that the Havana Syndrome will be solved by any official government body.

Armin Krishnan holds a PhD in Security Studies and an MA in Intelligence and International Relations and another MA in Political Science. His research focuses on emerging trends in warfare such as privatization, autonomous weapons, military neuroscience, gray zone conflict, and military ethics. He is currently the Director of Security Studies at East Carolina University, where he teaches courses on American foreign policy, intelligence studies, and international security.

---

113 The Daily Beast, "Havana Syndrome Might Be Real and It's Scary as Hell," *TDB*, February 2. 2023, available at: <https://www.thedailybeast.com/havana-syndrome-might-be-real-and-its-scary-as-hell?ref=scroll>.

Bibliography

Abbott, Daniel H. (ed.), *The Handbook of 5GW: A Fifth Generation of Warfare?* (Ann Arbor, MI: Nimble Books LLC, 2010).

Anderson, J., '"Brainwash" attempt by Russians?,' *Washington Post*, May 10, 1972, B15.

Arquilla, John and Ronfeldt, David, *In Athena's Camp: Preparing for Conflict in the Information Age* (Santa Monica, CA: RAND, 1997).

Atwood, Kylie, Lillis, Katie Bo, and Hansler, Jennifer, 'Biden Administration to Compensate Some "Havana Syndrome" Victims Up to $187,000,' *CNN*, June 24, 2022, available at: <https://www.cnn.com/2022/06/23/politics/havana-syndrome-victims-compensation/index.html>.

Baloh, Robert W. and Batholomew, Robert E., *Havana Syndrome: Mass Pyschogenic Illness and the Real Story Behind the Embassy Mystery and Hysteria* (Cham, Switzerland: Copernicus, 2020).

Barnes, Julian, "C.I.A. Officer Suffers Havana Symptoms," *New York Times*, September 20. 2021.

Barnes, Julian, "Report Criticizes C.I.A. Handling of Havana Syndrome Cases," *New York Times*, October 14, 2022.

Beeston, R., 'Russian Micro-waves Plan "to Drive U.S. Envoys Mad,"' *London Daily Telegraph*, May 11, 1972.

Belpomme, D. and Irigaray, P., "Electrohypersensitivity as a Newly Identified and Characterized Neurologic and Pathologic Disorder: How to Diagnose, Treat, and Prevent It," *International Journal of Molecular Sciences* 21, no. 1915 (2020): 1-20. doi:10.3390/ijms21061915.

Blank, Stephen, "Russian Information Warfare as Domestic Counterinsurgency," *The Journal of the National Committee on American Foreign Policy* 35, no. 1 (2013): 31-34.

Borger, Julian, 'Havana Syndrome has "Dramatically Hurt" Morale, US Diplomats Say,' *The Guardian*, February 10, 2022, available at: <https://www.theguardian.com/us-news/2022/feb/10/havana-syndrome-cuba-us-diplomats-afsa>.

Bouchet, Nicolas, 'Russia's "Militarization" of Colour Revolutions,' *Policy Perspec-*

tives 4, no. 2 (2016): 2-3.

Broad, W., "Microwave Weapons Are Prime Suspect in Ills of U.S. Embassy Workers," *New York Times*, September 1, 2018.

Bunker, Robert, "Unrestricted Warfare: Review Essay I," *Small Wars and Insurgencies* 11, no. 1 (2000): 115.

Cao, Kathy, Glaister, Sean, Pena, Adriana, Rhee, Danbi, Rong, William, Rovalino, Alexander, Bishop, Sam, Khanna, Rohan, and Saini, Jatin, "Countering Cognitive Warfare: Awareness and Resilience," *NATO Website*, May 20. 2021, available at: <https://www.nato.int/docu/review/articles/2021/05/20/countering-cognitive-warfare-awareness-and-resilience/index.html>.

CDC, "Cuba Unexplained Events Investigation: Final Report," December 3, 2019, available at: <https://nsarchive.gwu.edu/documents/cdc-report-havana-syndrome-medical-mystery-remains-unresolved/CDC%20%2320200983DOS255%20Final%20Report.pdf>.

Chekinov, S.G., and Bogdanov, S.A., "The Nature and Content of a New-Generation War," *Military Thought* 10 (2013).

Cochrane, Emily, "Health Drove US Envoys Out of Cuba," *New York Times*, August 11, 2017.

Dilanian, Ken, "FBI Acknowledges Some Agents May Have Havana Syndrome Symptoms," *NBC News*, November 24, 2021, available at: <https://www.nbcnews.com/health/health-news/fbi-acknowledges-agents-may-havana-syndrome-symptoms-rcna6504>.

Dilanian, Ken, "'Havana Syndrome' Symptoms in Small Group Most Likely Caused by Directed Energy, Says U.S. Intel Panel of Experts,' *NBC News*, February 2, 2022, available at: <https://www.nbcnews.com/politics/national-security/havana-syndrome-symptoms-small-group-likely-caused-directed-energy-say-rcna14584>.

Dorsey, Steve, 'Pentagon Turns Focus to Cuba Health "Attacks" Amid New Findings on American Victims,' *CBS News*, September 12, 2018, available at: <https://www.cbsnews.com/news/pentagon-turns-focus-to-cuba-attacks/>.

Dutton, Jack, "Havana Syndrome Symptoms Have Been Reported in these Countries," *Newsweek*, September 21, 2021, available at: <https://www.newsweek.com/havana-syndrome-china-austria-russia-1631102>.

Dylan, Huw and Maguire, Thomas J., "Secret Intelligence and Public Diplomacy in the Ukraine War," *Survival* 64, no. 4 (2022): 33-74.

Elder, A. and Chou, C.K., "Auditory Response to Pulsed Radiofrequency Energy," *Bioelectromgnetics Supplement* 6 (2003): S.162-S.173. doi:10.1002/bem.10163.

Entous, Adam, "Stealth Mode: How the Havana Syndrome Spread to the White House," *The New Yorker* 97, no. 14 (May 24, 2021).

Erickson, A., "All the Theories What's Happening to the Diplomats in Cuba: At Least 21 Americans Have Reported a Bizarre Rash of Symptoms," *Washington Post Blogs*, September 30, 2017, available at: <https://www.washingtonpost.com/news/worldviews/wp/2017/09/30/all-the-theories-about-whats-happening-to-the-diplomats-in-cuba/>.

Essex, Jamey and Bowman, Joshua, "From the Green Zone to the Havana Syndrome: Making Geographic Sense of Rotationality and Hardship in Diplomacy," *Diplomatica* 4, no. 1 (2022): 74-99.

Evans, R. and Novak, R., "The Microwave Affair," *Washington Post*, March 8, 1976, A19.

Finnegan, Conor and Smith, Cindy, "CIA Says Foreign Actor May Be Behind Some Havana Syndrome Cases," *NBC News*, January 20, 2022, available at: <https://abcnews.go.com/Politics/cia-foreign-actor-havana-syndrome-cases/story?id=82376545>.

Frey, A.H., "Human Auditory System Response to Modulated Electromagnetic Frequency," *Journal of Applied Physiology* 17, no. 4 (1962): 689-692.

Galeotti, Mark, *The Weaponisation of Everything: A Field Guide to the New Way of War* (New Haven, CT: Yale University Press, 2022).

Gardiner, H., "25th Person at U.S. Embassy Is Mysteriously Sickened," *New York Times*, June 21, 2018, available at: <https://www.nytimes.com/2018/06/21/us/politics/us-diplomat-cuba-embassy-illness.html>.

Gazis, Olivia, 'State Department, CIA Establish Federal Payment Rules for "Havana Syndrome" Victims,' *CBS News*, June 24, 2022, available at: <https://www.cbsnews.com/news/havana-syndrome-state-department-cia-establish-federal-payment-rules-for-victims/>.

Goldsmith, J.R., "Balancing the Interests of Patients, Science and Employees: Case

Study of RF (Microwave) Exposure of US Embassy Staff in Eastern European Posts," *The Science of the Total Environment* 184 (1996): 83-89.

Hammes, Thomas, *The Sling and the Stone: On War in the 21ˢᵗ Century* (St. Paul, MN: Zenith Press, 2004).

Hammes, Thomas, "Fourth Generation Evolves, Fifth Emerges," *Military Review* (May-June 2007): 14-23.

Ignatius, David, 'Dealing with "Havana Syndrome" Is a Policymaker's Nightmare,' *Washington Post*, October 28. 2021.

Ioffe, J., "The Mystery of the Immaculate Concussion," *GQ Magazine*, October 20, 2020, available at: <https://www.gq.com/story/cia-investigation-and-russian-mi crowave-attacks>.

Jacobsen, A., *Phenomena: The Secret History of the U.S. Government's Investigation Into Extrasensory Perception and Psychokinesis* (New York: Little Brown and Company, 2017).

JASON, "Acoustic Signals and Physiological Effects on U.S. Diplomats on Cuba," MITRE Corporation, November 2018.

Kaldor, Mary, *New and Old Wars: Organized Violence in a Global Era* (Cambridge, UK: Polity, 2001).

Kelemen, Michele and Shivaram, Deepa, 'VP Harris' Flight Delayed After Possible "Havana Syndrome" Incident in Hanoi,' *NPR*, August 24, 2021, available at: <https://www.npr.org/2021/08/24/1030663913/kamala-harris-havana-syndrome-vietnam-delay-embassy>.

Koffler, Rebekah, *Putin's Playbook: Russia's Secret Plan to Defeat America* (Washington. DC: Regnery Gateway, 2021).

Kolbe, Paul, Polymeropoulus, Marc, and Sipher, John, "Havana and the Global Hunt for U.S. Officers," *The Cipher Brief*, October 24, 2021, available at: <https://www.thecipherbrief.com/havana-syndrome>.

Kols, Richards, "NATO Must Meet Russia's Hybrid Warfare Challenge," *Atlantic Council*, July 3, 2018, <https://www.atlanticcouncil.org/blogs/new-atlanticist/nato-must-meet-russia-s-hybrid-warfare-challenge/>.

Levgold, Robert, *Return to the Cold War* (Cambridge, UK: Polity, 2016).

Liang, Qiao and Xiansui, Wang, *Unrestricted Warfare* (Beijing: PLA Literature and Arts Publishing House, 1999).

Libicki, Martin, "The Specter of Non-Obvious Warfare," *Strategic Studies Quarterly* (Fall 2012): 88-101.

Lind, William, Nightingale, Keith, Schmitt, John, Sutton, Joseph, and Wilson, Gary, "The Changing Face of War: Into the Fourth Generation," *Marine Corps Gazette*, March 2016, 86-90.

Lucas, Edward, *The New Cold War: Putin's Russia and the Threat to the West* (New York: St. Martin's Press, 2009).

McCreight, Robert, "Neuro-Cognitive Warfare: Inflicting Strategic Impact Via Non-Kinetic Effect," *Small Wars Journal*, September 16, 2022.

Meyers, S.L. and Perlez, J., "U.S. Diplomats Evacuated in China as Medical Mystery Grows," *New York Times*, June 6, 2018.

Michael, George, *Lone Wolf Terror and the Rise of Leaderless Resistance* (Nashville, TN: Vanderbilt University Press, 2012).

National Academy of Sciences, "New Report Assesses Illnesses Among U.S. Government Personnel and Their Families at Overseas Embassies," *News Release*, December 5, 2020, available at: <https://www.nationalacademies.org/news/2020/12/new-report-assesses-illnesses-among-us-government-personnel-and-their-families-at-overseas-embassies>.

New York Times, "Embassy Radiation Is Cut in Moscow," *New York Times*, April 26, 1976, 5.

Pollack, H., "Epidemiologic Data on American Personnel in the Moscow Embassy," *Symposium on Health Aspects of Non-Ionizing Radiation*, New York, April 9-10, 1979.

Power, Sean and Miner, Michael, "Report – Havana Syndrome: American Officials Under Attack," *Belfer Center Harvard University*, November 4, 2021, available at: <https://www.belfercenter.org/publication/report-havana-syndrome-american-officials-under-attack>.

Reed, Donald J., "Beyond the War on Terror: Into the Fifth Generation of War and Conflict," *Studies in Conflict & Terrorism* 31, no. 8 (2008): 684-722.

Relman and Pavlin (eds.), *An Assessment of Illness in U.S. Government Employees and Their Families in Overseas Embassies* (Washington, DC: National Academies Press, 2020).

Reppert, B., "Zapping an Embassy: 35 Years Later, the Mystery Lingers," *Times Daily*, May 22, 1988, 6D.

Reuters, "'Havana syndrome' U.S. Diplomats Get Benefits in Spending Bill," *Reuters*, December 16, 2019, available at: <https://www.reuters.com/article/us-usa-budget-congress-diplomats/havana-syndrome-u-s-diplomats-get-benefits-in-spending-bill-idUSKBN1YK24Z>.

Robles, F. and Semple, K., 'U.S. and Cuba Baffled by "Health Attacks" on American Envoys in Havana,' *New York Times*, August 12, 2017, A6.

Rubin, Rita, "More Questions Raised by Concussion-like Symptoms Found in US Diplomats Who Served in Havana." *JAMA* 319, no. 11 (2018): 1079-1080.

Schamiloglu, Edl, "Experts Believe US Embassies Were Hit With High-Power Microwaves – Here's How the Weapons Work," *SciTechDaily*, February 12, 2022, available at: <https://scitechdaily.com/experts-believe-us-embassies-were-hit-with-high-power-microwaves-heres-how-the-weapons-work/>.

Schumaker, James, "Before Havana Syndrome, There Was Moscow Signal," *Foreign Service Journal* (January/ February 2022), available at: <https://afsa.org/havana-syndrome-there-was-moscow-signal>.

Swanson, Randal L., Hampton, Stephen, Green-McKenzie, Judith, Diaz-Arrastia, Ramon, Grady, Sean, Verma, Ragini, Biester, Rosette, Duda, Diana, Wolf, Ronald L., and Smith, Douglas H., "Neurological Manifestations Among US Government Personnel Reporting Directional Audible and Sensory Phenomena in Havana, Cuba," *JAMA* 319, no. 11 (2019): 1125-1133.

The Daily Beast, "Havana Syndrome Might Be Real and It's Scary as Hell," *TDB*, February 2. 2023, available at: <https://www.thedailybeast.com/havana-syndrome-might-be-real-and-its-scary-as-hell?ref=scroll>.

Thomas, Timothy, "The Mind Has No Firewall," *Parameters* (Spring 1998): 84-92.

U.S. Congress, H.R.4914 passed on August 3, 2021, available at: <https://www.congress.gov/117/bills/hr4914/BILLS-117hr4914ih.pdf>, Section 2 (13).

U.S. Department of Defense, "Bioeffects of Selected Nonlethal Weapons," *Depart-*

ment of the Army (1998), declassified on December 6, 2006.

U.S. Department of Defense, "Directed Energy Futures 2060: Visions for the Next 40 Years of U.S. Department of Defense Directed Energy Technologies," (2021), available at: <https://www.afrl.af.mil/Portals/90/Documents/RD/Directe d_Energy_Futures_2060_Final29June21_with_clearance_number.pdf?ver =EZ4QY5MG5UK2LDdwiuPc6Q%3D%3D>.

U.S. Department of State, "Havana Cuba: Accountability Review Board," June 2018, Declassified on October 1, 2019.

U.S. Office of the Director of National Intelligence, "IC Expert Panel on Anomalous Health Incidents Report: Executive Summary," declassified February 1, 2022.

Van Creveld, Martin, *The Transformation of War* (New York: Simon & Schuster, 1991).

Verma, R., Swanson, R.L., Parker, D., Ould Ismail, A.A., Shinohara, R.T., Alapatt, J.A., Doshi, J., "Neuroimaging Findings in U.S. Government Personnel with Possible Exposure to Directional Phenomena in Havana, Cuba," *Journal of the American Medical Association* 322, no. 4 (2019): 336-347. doi:10.1001/jama.2019.9269.

Ward, Alexander and Desiderio, Andrew, 'U.S. Foe or Specific Weapon Not Behind Sustained, Global Campaign Causing "Havana Syndrome," CIA Finds,' *Politico*, January 19, 2022, available at: <https://www.politico.com/news/2022/01/19/havana-syndrome-cia-causes-527457>.

Wilkinson, T., "Did Covert Spy Gear Cause U.S. and Canadian Diplomats in Cuba to Lose Their Hearing?," *Los Angeles Times*, August 10, 2017, available at: <https://www.latimes.com/nation/la-fg-cuba-hearing-20170810-story.html>.

Double Agent Snow: The Beginnings of the Double Cross Network

Mona Parra, PhD

Grenoble Alpes University – ILCEA4
mona.parra@univ-grenoble-alpes.fr

ABSTRACT

Double agent SNOW, who was described by British intelligence officers as an "underfed rat", did not seem destined to play a significant part in Britain's World War Two strategy. Before and during the war, he worked for both German and British intelligence. Where his true loyalty lay remains, to this day, unclear. He was the first member of "Double Cross", a network which would go on to include more than 100 double agents under British control. The lack of compartmentalisation meant that this dubious agent could jeopardise many others. Despite his duplicity, he allowed British officials to discover the strengths and weaknesses of the German intelligence services and of their own agencies. He was a true asset who helped achieve cryptanalytic breakthroughs and who contributed to foiling numerous espionage attempts on British soil. Above all, his handling was a cornerstone for British intelligence in that it taught case officers how to run a double agent, the risks entailed, and the benefits they could draw from it.

L'agent double SNOW, décrit par les officiers traitants britannique comme « un rat sous-alimenté », ne semblait pas destiné à jouer un rôle significatif dans la stratégie de le Seconde Guerre mondiale. Avant et pendant la guerre, il travailla pour le renseignement allemand et britannique. Sa véritable allégeance demeure, de nos jours, incertaine. Il fut le premier membre du réseau « Double Cross » qui comprendrait plus de cent agents sous contrôle britannique. Le manque de cloisonnement signifiait que tout agent pouvait compromettre nombre de ses collègues. Malgré sa duplicité, il permit aux Britanniques de découvrir les forces et les faiblesses des services allemands de renseignement et de leurs propres agences. Il fut un véritable atout qui permit de réaliser des percées cryptanalytiques et qui contribua à déjouer de nombreuses tentatives d'espionnage sur le sol britannique. Surtout, cette expérience fut un élément clé pour le renseignement britannique car il permit à ses officiers traitants d'apprendre comment superviser un agent double, les risques encourus, et les avantages qu'ils pouvaient en tirer.

doi: 10.18278/gsis.8.1.6

Keywords: Agent Snow, espionage, HUMINT, MI5, World War Two, WWII

El doble agente Snow: los inicios de la red Double Cross

Resumen

A finales de octubre de 1939, tras una entrevista con un agente doble anónimo, un oficial de inteligencia británico escribió el siguiente informe sobre sus rivales alemanes: "[El nombre del agente se elimina para preservar su anonimato] dice que se referían a Snow como su hombre número uno. en Inglaterra. (Si esto no es un farol, lo que parece probable, ¡entonces están bastante mal!)". Este pasaje escrito al comienzo de la guerra da fe del fracaso de la inteligencia alemana y es una indicación de los problemas que continuarían durante todo el conflicto; Por el contrario, los británicos lograron sus primeros éxitos en este campo ese mismo año.

Snow fue un súbdito canadiense nacido en Gales. Regresó al Reino Unido a principios de los años 1930. Como ingeniero eléctrico, viajaba regularmente a Europa en viajes de negocios. Reconociendo el valor estratégico de tales viajes, se acercó repetidamente a las agencias de inteligencia británicas para ofrecer sus servicios en los años previos a la guerra. Fue rechazado una y otra vez pero aun así persistió. Esta renuencia británica a reclutarlo se debió al descubrimiento de que Snow también trabajaba con los alemanes. Sin embargo, los británicos no lo detuvieron hasta la declaración de guerra. A partir de septiembre de 1939 se vio obligado a comunicarse con los alemanes bajo control británico. Operó de esta manera entre septiembre de 1939 y marzo de 1941. A lo largo de este período, las sospechas británicas sobre la verdadera lealtad de Snow se reavivaron con frecuencia. Sus dudas culminaron en marzo y abril de 1941, lo que les llevó a poner fin a sus actividades. Pasó el resto de la guerra en la cárcel.

Palabras clave: Agente Snow, espionaje, HUMINT, MI6

双重特工Snow：背叛网络的开端

摘要

1939年10月下旬，一位英国情报官员在采访一位未透露姓名的双重间谍后，写下了关于他的德国竞争对手的报告："该特工（名字被删除以保持匿名）说他们将Snow称为他们在英国的头号人物。（如果这不是虚张声势——虽然很可能是——那么他们的处境就相当糟糕了！）"。这段在战争开始时写下的段落证明了德国情报的失败，并表明了在整个冲突期间将持续存在的问题；相比之下，英国人于同年在该领域取得了首次成功。

Snow是出生于威尔士的加拿大人。他于1930年代初返回英国。作为一名电气工程师，他定期前往欧洲出差。认识到这类旅行的战略价值，他在战争开始之前的几年里多次与英国情报机构联系，希望为其提供服务。虽然他一次又一次地被拒绝，但仍然坚持了下来。英国人不愿意招募他是因为发现Snow也与德国人合作。然而，直到宣战时，英国人才阻止他。从1939年9月起，他被迫在英国的控制下与德国人联系。1939年9月至1941年3月期间，他就以这种方式进行操作。在此期间，英国方面经常怀疑Snow的真实效忠。这种怀疑在1941年3月和4月达到顶点，最终终止了Snow的活动。他在监狱里度过了接下来的战争时期。

关键词：Snow特工，间谍活动，人力情报，军情六处

I n late October 1939, following an interview with an unnamed double agent, a British intelligence officer wrote the following report on his German rivals: "[The agent's name is deleted to preserve his anonymity] says they referred to Snow as their No. 1 man in England. (If this isn't bluff, which seems likely, then they are pretty badly off !) [1]". This passage written at the start of the war attests to the failure of German intelligence, and is an indication of problems that would continue throughout the conflict; in contrast, the British achieved their first successes in this field in the same year.

Snow[2] was a Canadian subject born in Wales. He returned to the UK in the early 1930s. As an electrical engineer, he regularly travelled to Europe for business trips. Recognising the strategic value of such travel, he repeatedly approached British intelligence agencies to offer his services in the years

---

1   The National Archives, Kew, London (henceforth NA), NA/KV/2/468, report dated 20 October 1939. The rest of the account of the discussion with this anonymous agent provides several elements suggesting that this person is probably Gwilym Williams, called "G.W." by the British services.

2   The name Snow is very frequently capitalised in MI5 files, and I will adopt this convention in the article.

leading up to the war. He was turned down again and again but nevertheless persisted. This British reluctance to recruit him was due to the discovery that Snow also worked with the Germans. However, the British did not stop him until the declaration of war. From September 1939 he was forced to communicate with the Germans under British control. He operated in this way between September 1939 and March 1941. Throughout this period, British suspicions about Snow's true allegiance were frequently reignited. Their doubts culminated in March and April 1941, which led them to put an end to his activities. He spent the rest of the war in jail.

Despite all the doubts about Snow's sincerity, he holds a unique place in the history of British espionage. He was the first agent in the Double Cross network of double agents which was set up during the Second World War. This British network included more than a hundred double agents. During the war, Double Cross was able to foil espionage projects on British soil, to collect information on German intelligence and to "feed" the Germans with disinformation, culminating in Operation Fortitude, which convinced them that the landings would take place in the Pas-de-Calais and not in Normandy. These agents played a central role in the allied strategy. Despite the immense risks, relying on Snow had a triple benefit for the British. It allowed them to gather information on Germany, to better understand the spy network of the

Third Reich, and also to draw up rules for dealing with double agents. We can therefore ask how the risky collaboration with Snow allowed the British to build an invaluable network of double agents.

In a first part we will study the way in which Snow was recruited before analysing his activities under British control. This will then allow us to address his crucial role in the genesis of Double Cross.

Snow, whose real name was Arthur Graham Owens, was born in Wales in 1899. His business trips frequently led him to German ports, which he supplied with submarine batteries. This is why the British Admiralty approached him in the mid 1930s[3] . This is the first step in the agent acquisition process: "spotting". At the end of 1936, the British services quickly realised that Snow was also providing information to Germany, at a time when tensions between the two countries were escalating. They announced to Owens that they were ending all contact with him. Inevitably, his talks with them allowed him to learn details about the British services that could be damaging to them. The British judged that the potential gain was not worth the risk. The files on Snow, which were only declassified in 2001, are full of evidence to this effect.

In April 1938, Lieutenant-Colonel Edward Hinchley-Cooke (MI5) reminded him of a September 1937 warning that British intelligence re-

---

3   Hayward, James, *Double Agent Snow. The True Story of Arthur Owens, Hitler's Chief Spy in England*, London, Simon and Schuster, 2013, locations 221-227 (Kindle edition).

fused to have any contact with him[4]. A few months later, on 24 September 1938, Hinchley-Cooke, now a Colonel, interrogated Snow and reiterated this warning. He warned him that his words could be used against him if he was sued for espionage[5].

There are many reasons for this lack of confidence in Snow. His answers were evasive, to say the least, and frequently contradictory, which made it very difficult to trust him[6]. He was portrayed as a "bad lot, requiring very discreet handling"[7]. Snow is "about 5 feet tall and of thin build"[8]. One file describes him as "the typical Welsh 'underfed' type, very short, bony face, ill-shaped ears"[9]. The British also used the term "underfed rat"[10], the word "rat" having of course a strong negative connotation. Snow is presented as a shady individual. This judgement is perhaps partly coloured by the ancestral rivalry between the English and the Welsh.

However, throughout this period, Snow provided valuable information to the British services which nevertheless frequently insisted he should

stop contacting them, or at least refused to acknowledge their association with him. Snow revealed the identity of members of the German secret service, as well as the addresses used for communications. The British officers took care to meticulously record all these names and contact details in the hope of getting a fairly clear picture of the German organisation, as evidenced by the multiple annotations on the interrogation reports[11].

It was through the classification procedure established by MI5 that Snow's double-dealing was initially brought to the attention of the British. Snow sent his correspondence to an address in Germany known to the British. A pre-war double agent, Christopher Draper, had informed them that it was used by the Abwehr, Germany's most notorious military intelligence service[12]. These suspicions about Snow were corroborated by Snow's wife. Shortly before the war, after he left her, she decided to reveal his activities to the authorities[13].

She claimed that he was very ac-

---

4   NA/KV/2/445, note dated 7 April 1938.

5   NA/KV/2/445, transcript of an interrogation of Snow by Colonel Hinchley-Cooke in the presence of two other persons whose names are deleted, on 24 September 1938 at Scotland House.

6   NA/KV/2/445, report entitled "Snow. Report of Interview with D.E.E.", dated mid-March 1938.

7   NA/KV/2/446, letter to Major Boyle, 2 February 1939.

8   NA/KV/2/445, note dated 5 September 1938 entitled "re Snow".

9   Andrew, Christopher, *Defend the Realm. The Authorized History of MI5* (New York: Alfred A. Knopf, 2009), p. 212, based on NA/KV/2/444.

10 NA/KV/2/445, note dated 5 September 1938 entitled "re Snow".

11 NA/KV/2/445, Snow's interrogation, 24 September 1938.

12 Andrew, *op. cit.* p. 212.

13 NA/KV/2/446, report of Mrs Snow's testimony (18 August 1939) to the Special Branch of the Metropolitan Police.

tively helping the Germans. He tried to force several of their relatives to provide information to the secret agencies of the Reich, and even threatened to kill her if she revealed his activities to the police[14]. He used his profession as a cover for his illegal activities, for example referring to "samples" to discuss the information he sent to Germany[15]. His wife testified that he gave them details of British airfields. She added that she had destroyed codes used by the Royal Air Force so that he could not pass them on to the Germans, but that he may still have some[16].

The British studied the motives behind Snow's ambiguous stance. The assessment of Snow's trustworthiness, a key element in the recruitment cycle, was carried out repeatedly in order to understand how best to use this individual. Owens explained that his resentment of the UK stemmed from the fact that he, his half-brother and his father had designed a shell to shoot down German Zeppelins during the First World War. He claimed that the British authorities did not admit that he and his relatives were the inventors, and that his family had lost hundreds of thousands of pounds as a result[17]. The money the Germans paid him for his

services largely explains the ambivalent attitude of this frequently cash-strapped man[18]. His motivations seem to be in line with the MICE model listing the reasons why a spy serves an intelligence agency (Money, Ideology, Coercion or Compromise and Ego or Excitement), a model which has been questioned[19] but which provides a relevant theoretical framework here. Money and ego prevailed in his relationship with the Germans, whereas when working with the British he probably sought the excitement associated with this clandestine activity and wanted to put himself forward. On the other hand, coercion soon became the rule in his relationship with the British authorities, who forced him to continue working for them. At no time did ideology come into play, a factor which is the best guarantee of a source's reliability. Snow was never seen as a reliable asset. The first stage of Snow's activities came to an end in September 1939, when Snow's operations reached unprecedented levels, making him the first British double agent of the Second World War.

On 3 September 1939, the UK declared war on Germany. The secret services had decided that Snow would be imprisoned for espionage. He contact-

---

14 *Ibid.*

15 NA/KV/2/445, letter from Snow (who signs "Johnny", the nickname the Germans gave him) to 'Dr. Rantzau', his German handler, 7 October 1937.

16 NA/KV/2/446, report of Mrs Snow's testimony (18 August 1939) to the Special Branch of the Metropolitan Police.

17 NA/KV/2/446, letter to Hinchley-Cooke, 24 March 1939.

18 West, Nigel, Roberts, Madoc, Snow. *The Double Life of a World War II Spy*, London, Biteback, 2011, locations 233 and 362 (Kindle edition).

19 Among others in Burkett, Randy, "An Alternative Framework for Agent Recruitment: From MICE to RASCLS", Studies in Intelligence, vol. 57, no. 1, pp.7-17, 2013.

ed them and a meeting was organised on September 4th at Waterloo station. Snow wanted to give further information to his British contacts, but they immediately arrested him and he was sent to Wandsworth Prison[20] . The house where he was staying was searched and a radio transmitter was found. It had been buried in the garden to conceal its presence[21] . The British thought that Snow could be an asset and, despite the risks, they finally decided to treat him as a double agent in his own right, and to allow him to communicate with the Germans as if he were not under British control.

The first test took place on 8 September. On this occasion, Snow blew a fuse while manipulating the transmitter. The next day, the British finally managed to send their first message : "All ready. Have repaired radio. Send instructions. Now awaiting reply"[22] . Snow gave contradictory answers to his British interrogators about the procedure for communicating with the Germans, which fuelled their suspicions[23] . He was initially criticised for his lack of cooperation, but after some veiled threats he seemed to

change his attitude[24] . On 11 September, he was transferred to Kingston police station. The archival record of this move contains Snow's real surname, which the staff preparing the documents for declassification normally erase. His surname, Owens, is therefore revealed[25] . The code name assigned to him by the British, Snow, is a partial anagram of his surname. Gradually, the British gained confidence in Snow, and in mid-September they decided to allow him to travel to Holland to meet his German contacts[26] . However Snow and his lover, Lily, were under surveillance[27].

An incident almost put a premature end to Snow's activities. In May 1940, he was invited to rendezvous with his Abwehr handler, Ritter, in the North Sea. Owens was accompanied by a new recruit, "BISCUIT". During the trawler trip, the two men became increasingly suspicious of each other. In this climate of suspicion, BISCUIT panicked, imprisoned Snow, and ordered the boat to turn back. To justify their absence at the agreed place and date, they gave the excuse of meteorological difficulties[28] . The Germans expressed some doubts

---

20   NA/KV/2/446, letter from the Special Branch of the Metropolitan Police, 6 September 1939.

21   *Ibid.*

22   West, Nigel, Roberts, Madoc, *op. cit,* location 818.

23   *Ibid,* location 818.

24   NA/KV/2/446, report on "Snow", dated early September 1939.

25   NA/KV/2/446, report on "Snow" by T.A. Robertson, 14 September 1939.

26   *Ibid.*

27   *Ibid.*

28   Andrew, Christopher, *op. cit,* pp. 248-249. Masterman, John Cecil, *The Double Cross System in the War of 1939 to 1945,* New Haven and London, Yale University Press, 1972, pp.42-44. J.C. Masterman played an active role in the control of Snow. He became the chairman of the Twenty

but finally seemed convinced. This episode illustrates the difficulties inherent in the cooperation of several double agents. However, such associations are common in the case of Snow, which is contrary to the elementary rules of security for spies. Shortly afterwards, another trip with a double agent dealt a fatal blow to Snow's activities. In each case, the agent's true allegiance, and suspicions about it, led to potentially tragic developments.

In October 1940, in order to ensure that Snow could continue to communicate with the Germans, his handlers decided he should leave London, in the context of the *blitz*. Rather cynically, J.H. Marriott, the officer who suggested this move, explained that it would prevent "Snow's wireless transmitter [from] being destroyed in an air raid"[29].

Although Snow was a crucial source of information, doubts about his honesty were never fully resolved. Regularly, new incidents alerted his British handlers. A letter to Robertson, Snow's handler, announced a final twist on 21 March 1941 : the Germans had discovered that Snow was operating under British control, which he confessed to when they told him of their suspicions in Lisbon[30]:

Committee established to supervise double agents from January 1941.

29 NA/KV/2/449, letter from J.H. Marriott, 1 October 1940.

30 NA/KV/2/449, letter to T.A. Robertson, 21 March 1941.

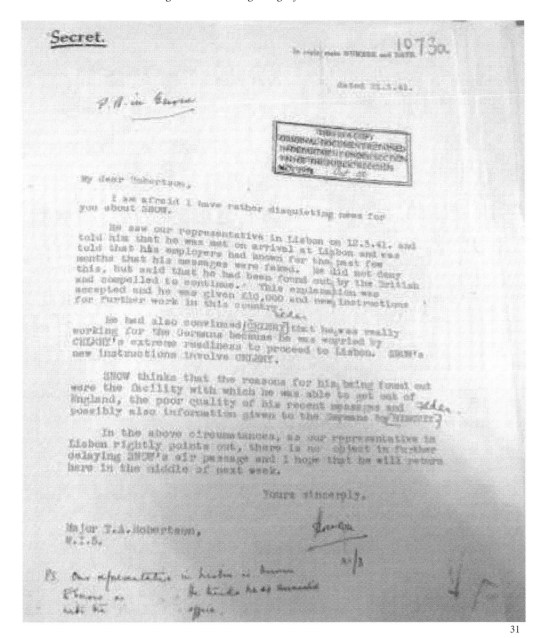

31

On his return to the UK after his fateful stay in Portugal, Snow was subjected to multiple interrogations, during which MI5 officers repeatedly asked him to explain the exact circumstances in which the Germans revealed to him that they knew of his connection to the British. His entire stay was the subject of extensive questioning, to which Snow answered at length, but in a confused and contradictory manner. The British did not know if they could trust him. Many things remained unclear despite the long interviews. The British were not even sure this confrontation had actually taken

---

31   NA/KV/2/449, letter to T.A. Robertson, 21 March 1941.

place. Indeed, Snow claimed to have warned "Celery", another double agent who was with him in Lisbon, that the Germans knew about his collaboration with the British. But the details provided by Snow as to the circumstances of this discussion were very vague, and he never gave the same answer to his interrogators. Celery denied that Snow had warned him that the Germans knew about this double-dealing[32]. Celery and Snow constantly accused the other of treason. This caused deep concern for the British. All the other double agents who had been in contact with them were potentially compromised. Snow did not give a clear answer to help determine if the Germans suspected the other double agents. If the British asked them to "feed" the Germans with disinformation about the strategy of the Allies, this might backfire on the Allies at a time when a German invasion of Britain was feared. There is much evidence to suggest that Snow's account of this Abwehr interview was unfounded. How else can we explain that, without extensive interrogation, the Germans agreed to believe him, and even entrusted him with 10,000 pounds and sabotage material before he left Portugal[33]? MI5 officer J.H. Marriott concluded his account of one of Snow's interrogations with the following sentence, a clear indication of his annoyance: "I am more than ever

convinced that Snow's is a case not for the Security Service, but for a brain specialist"[34].

MI5 was unable to decide on Snow's future, and decided to call in a doctor to get an opinion on his medical condition, which he regularly complained about, but also on the patient's mental disposition. The clinician took Snow for an X-ray to see if his ailments were real. The result was clear: there seemed to be no medical ground for Snow's complaints, but more importantly, the doctor's verdict was that Snow "[...] was a consummate liar and that his word could not be trusted in anything that he said [...] he would try to deceive us in any way that suited his purpose"[35]. This judgement seems to go beyond a strictly medical framework. After these lengthy investigations, MI 5 decided to incarcerate Snow. On 10 April 1941, T.A. Robertson and his colleague J.C. Masterman asked him to send a message to the Germans, explaining that his psychological state and his health forced him to stop his activity[36]. Coded German radio communications, in particular a message sent by Ritter and deciphered by the British, revealed the German authorities' disbelief in Snow's sudden illness, and proved that they suspected he was not telling them the whole truth[37].

---

32 NA/KV/2/450, interrogation of Snow and Celery.

33 NA/KV/2/450, interrogation of Snow, 31 March 1941.

34 NA/KV/2/450, "Note on the Interrogation of Snow" by J.H. Marriott, 3 April 1941.

35 NA/KV/2/450, letter dated 18 April 1941.

36 NA/KV/2/450, interrogation of Snow by Major Robertson and Captain Masterman on 10 April 1941.

37 NA/KV/2/450, transcript of a German communication between Cyrenaica and Hamburg dated 1

A report on Snow dated 31 July 1941 gives yet another interpretation of what happened. In it, Snow's German handler Ritter stated that on arriving in Lisbon Snow had doubts about the reliability of Celery, whom he suspected of being a British intelligence agent[38]. Snow was finally put in prison on 23 April 1941[39]. However, the twists and turns did not stop there. In August 1941, MI5 learned from a fellow prisoner that Snow was trying to escape, and that his preparations were well under way thanks to the complicity of one of the guards[40]. Measures were taken to thwart these plans. When Ritter fell into Allied hands after the German surrender, he was interrogated and stated in 1946 that he knew from the events in Lisbon that Snow was serving the British but took no action[41]. This would confirm Snow's account. However, Ritter's son-in-law maintained after his death that these confessions were probably false and intended to portray him in a favourable light to the victors in order to secure him lenient treatment[42]. None of the three main protagonists of this incident offers a definitive authoritative version.

Throughout Snow's period of operation, the UK's military prospects were highly uncertain. For example, in early 1941, its inhabitants still feared that their island would be invaded by the Germans[43]. The British planners therefore defined procedures to be put in place in this eventuality, under the code name "Hegira". They decided that should this situation arise, an MI5 officer, Mr Marriott, would take "Mrs Snow and the baby" to Wales. "Mrs Snow" actually referred to Snow's lover Lily Bade, not his wife. Snow's adult son from his first marriage would also be taken to Wales. He was aware of his father's activities. All incriminating documents would be destroyed and the transmitter would be taken to Wales in the hope of continuing communications with the Germans. In the event of their capture by the enemy, Marriott and his colleague Reed were instructed to take "all necessary steps" to ensure that Lily and Snow's son did not fall into German hands[44]. Similar instructions were given for other double agents. Christopher Andrew, one of the UK's leading intelligence experts who has had exceptional access to classified material, points out that this is the only known occasion when a director of MI5

June 1941. The page is marked, "Top Secret U. Snow", "U" being a symbol for "Ultra", the result of the British decryption of the most elaborate Axis codes. "3504" is the code name assigned to Snow by the Abwehr.

38 West, Nigel, Roberts, Madoc, *op. cit*, location 3709.

39 NA/KV/2/450, "note on the case of Snow Junior", 19 August 1941.

40 NA/KV/2/450, report by Captain J.C Masterman, 14 August 1941.

41 West, Nigel, Roberts, Madoc, *op. cit*, location 4041.

42 *Ibid,* location 4103.

43 Andrew, Christopher, *op. cit.* p. 257.

44 NA/KV/2/450, "Orders in the Event of an Invasion", for Mr Marriott. This typed document bears a handwritten date: 3 April 1941.

has condoned such executions; this decision was based on the belief that if a double agent tried to help the invading army, he or she could be seen as enemy combatants[45] . The invasion never took place, so these instructions were never followed. These orders concerning Lily were reiterated at the end of 1941[46] . Summary executions, without trial, of individuals whose guilt was not proven (in the case of Lily Bade) were considered.

The problems raised by Snow's activities allowed British officers to discover how to ensure that such operations were successful and to draw up a certain number of rules for carrying them out. These rules then guided them in their handling of the other double agents of the *Double Cross* network.

The first rule was to remain suspicious at all times because a double agent is likely to change sides. Vigilance is the order of the day. In December 1939, Snow was warned that his flat would be under surveillance, as well as his movements, which he accepted. He did not mention the reasons why the British were taking this step, but added that this would enable them to obtain information on his German contacts[47] . He had often been watched on previous occasions, as in October 1939[48].

Another principle which was often emphasised, by both the British and the Germans, is to ensure that contacts between double agents are kept to a minimum. Each should only know a few details about the others[49] . The files in the Snow archives show that this rule was hardly followed. More than a dozen other double agents are regularly mentioned alongside Snow in the MI5 archives. To protect Snow and TATE, another double agent, the British decided to "eliminate" a third spy called SUMMER, making the Germans believe that he had not had time to reveal any information to the British[50]. He was arrested and ceased all exchanges with the Germans[51].

Finally, another example of the principles which were discovered or rediscovered through Snow's handling it that it is important to ensure that a spy leads the life he would have if he had not switched sides. For example, British officers asked Snow to travel to collect information requested by the Germans, even when it was given to him by his British handlers. This was a security measure to ensure that he was prepared to give convincing answers to any questions about his intelligence sources. He

---

45   Andrew, Christopher, *op. cit.* pp. 258-259.

46   NA/KV/4/211, "Orders in the Event of an Invasion", for Mr Marriott. This typed document bears a handwritten date: 3 December 1941.

47   NA/KV/2/446 letter dated 2 December 1939 with no signature.

48   NA/KV/2/446, report dated 25 October 1939, on the watch on Snow organised by the Special Branch of the Metropolitan Police.

49   NA/KV/2/449, report dated 15 January 1941 signed by J.C. Masterman.

50   *Ibid*.

51   Masterman, John Cecil, *op. cit.* pp. 50-52.

went near several airfields; however, the information he obtained was subject to approval by the British authorities before being passed on to the Germans[52].

Snow's career as a double agent raised many questions among MI5 officers. In order not to reveal to the Germans that he was under British control, Snow had to provide them with information as if he were operating normally in the UK. This information could be risky from a military point of view, with potential loss of life. For example, Snow's German handlers asked him for weather information. Nowadays it is hard to imagine that a lot of data that is immediately available to us, thanks to the media and the Internet, was difficult to obtain at the time. Weather conditions belong to this category. It may seem irrelevant in a military context, but in fact it is crucial information which is necessary to direct aircraft and missiles towards their targets and to determine whether they can fly at all. Snow received detailed questionnaires on everything from meteorology to the composition of bomber squadrons[53]. The dilemma was even greater when the Germans asked for data on a specific city in preparation for an air attack. How could it be protected without compromising the agent? What answer should be given?

For the sake of credibility, it was decided that the meteorological information to be provided to Snow should not come from a meteorological expert, but rather from a simple observer. The files dating from October 1939 are proof that the details given to the Germans were not limited to this type of data. For example, his handlers prepared information on troop movements[54], as well as precise details of British airfields[55]. To be credible, the information provided by the different agents had to be consistent. Some data was considered too sensitive to be communicated to the Germans. Thus, in October 1940, it was decided that it was not possible to answer the Germans' questions about the bridges crossing the Royal Military Canal[56] in the south-east of England. It is however essential that an agent should provide enough credible or true information to be believed by his handlers without revealing that he has betrayed them. In December 1939, Snow was reprimanded by Nikolaus Ritter, his German handler, because the weather data he had provided was unsatisfactory[57].

The problems with Snow were recurrent. After a stay in Holland in September 1939, Snow went to Brussels in October. He came back with several detonators hidden in a piece of

---

52   NA/KV/2/446, report of an interview with Snow, no date or signature, dated mid-November 1939.

53   NA/KV/2/449, letter dated 5 October 1940.

54   NA/KV/2/446, note by T.A. Robertson dated 11 October 1939.

55   NA/KV/2/446, note by T.A. Robertson dated 15 October 1939.

56   NA/KV/2/449, note dated 1 October 1940, unsigned.

57   NA/KV/2/446, report of an interview with Snow on 20 December 1939.

wood, and 470 pounds[58] . He worked with another double agent, G.W. The Germans had indeed asked him to put them in contact with the Welsh nationalist movement, in order to organise sabotage on British soil. The Germans offered to train Snow and G.W. in sabotage and intelligence[59] . The handling of these two agents obviously created formidable dilemmas. How far could their British handlers go to establish their credibility?

It took a long time to establish the procedures for the control of agents. In December 1940 and January 1941, more than a year after Snow's activities had begun, measures were taken to ensure that the section that controlled double agents, MI5's B1a branch (which worked closely with MI6), could cooperate smoothly with the other branches which were involved. The Twenty Committee was set up in order to have the support of the highest authorities in British civilian and military intelligence. Its mission was as follows: through it, B1a could be allowed to give information to the Germans and the Directors of Intelligence sought its help to mislead the enemy[60] . The choice of the term "Twenty Committee" is linked to the name of the network, "Double Cross". In Roman numerals, the double cross, "XX", means twenty, which gave the commit-

tee its name. The first meeting was held in January 1941, and thereafter it met weekly. The issues to be dealt with were indeed both numerous and urgent. Masterman, the head of this operation, claimed there was "harmonious collaboration between a large number of different services and departments"[61].

Despite all the risks, the benefits were enormous. Snow gained invaluable information about the enemy's methods. He revealed the methods used by the German services to send instructions to their spies by reducing an entire page of text to a tiny dot using "microphotography" and then hiding this dot, for example, on the back of a stamp[62] , or within an innocuous-looking document. Through Snow, the British obtained valuable information about the codes that the Abwehr had asked him to use[63] . The pre-war interviews which the British services conducted with Snow also revealed deep flaws in the UK's preparation for a potential war and in the preparation of its secret services. Snow revealed that the Germans knew that "Major Kennedy" had helped British intelligence discover a spy. He revealed that the Germans had a great deal of information about British military forces[64] . Archival records show the interest of the officers in the information given by Snow when he

---

58   NA/KV/4/185, Guy Liddell's diary, entry for 26 October 1939.

59   NA/KV/4/185, Guy Liddell's diary, entry for 27 October 1939.

60   "Traffic and deception", note dated 21 July 1942, TNA KV 4/213.

61   Masterman, John Cecil, *op. cit*, p. 65.

62   NA/KV/4/185, diary of Guy Liddell, entries for 27 and 31 October 1939.

63   NA/KV/2/445, Snow interrogation, 24 September 1938.

64   NA/KV/2/445, report on an interview with Snow on 28 September 1938.

told them, in September 1938, that he would be notified of the first British towns to be bombed so that he could provide the Germans with the coordinates of the power stations and steelworks. A large handwritten line runs along this entire paragraph of the interrogation report, emphasising the attention that was paid to it. Such information could help limit the loss of life and property, by strengthening the defences of the intended targets, and by providing false data to the attackers. However, one had to avoid compromising one's agent. This proves how complex the handling of double agents was, as this document[65] shows.

After these eventful years, Snow spent the rest of the war behind bars. Although the handling of the first double agent of such magnitude was not without its flaws, the Snow case was a success story. It helped to arrest many Abwehr spies as soon as they arrived on British soil. Masterman states that after the fall of France in 1940, and until 1944, all spies sent by Germany fell into British hands, and a significant proportion of them came under British control[66]. This was an invaluable asset for the UK in the context of the war. The British feared that a Fifth Column operating on their territory would do serious harm to the country. British intelligence prevented this from happening. The control of Snow also led to sig-

nificant progress in decrypting German ciphers: messages sent by Snow to his handlers were then coded using the famous Enigma, which led to major cryptanalytical breakthroughs which were crucial to the Allies[67]. The process of trial and error in Snow's handling allowed the British to discover how to take advantage of their double agents throughout the Second World War. Some of those associated with Snow were able to continue their activities after his final arrest, like G.W. who continued to operate until February 1942[68]. This proves that MI5 was able to protect these very special agents as best it could. The Snow archives show that in every country, but especially in the UK, the secret services have always occupied a special place : the Snow files were only made public in 2001[69], more than fifty years after the end of the conflict. The redaction of declassified documents by the British authorities makes the work of researchers even more complicated: names are erased, portions of sentences and even entire pages are obliterated, leaving entire spaces empty, to preserve the secrecy of this world. Ironically, in 1972 MI5 and SIS even considered the imprisonment of J. C. Masterman, the former director of the Twenty Committee who played such a key role during the war, when he published *The Double-Cross System*, which shed light on the British double-agent network,

---

65 NA/KV/2/445, Snow interrogation, 24 September 1938.

66 Masterman, John Cecil, *op. cit*, p. 6.

67 West, Nigel, Roberts, Madoc, *op. cit*, location 845.

68 Masterman, John Cecil, *op. cit*, Appendix 1: "Double-Cross Agents in the United Kingdom".

69 As the catalogue of the National Archives of Kew proves: http://discovery.nationalarchives.gov.uk.

creating a real historiographical turning point. The Official Secrets Act, which silences anyone with a state secret, still holds sway.

Mona Parra, an ENS Cachan alumna, is an Associate Professor at Grenoble University and is a member of the ILCEA4 research team. Her 2015 PhD dissertation investigated the protection of British, American and German intelligence during the Second World War. Her research focuses on the British intelligence services, on their cooperation with their foreign counterparts, in particular within the UK-US special relationship, and on the connexion between intelligence and international relations.

Selected bibliography

Archives

The National Archives, Kew, London, Great Britain.

KV 2/445-KV 2/450

KV 2/468

KV 2/674

KV 4/1-KV 4/3

KV 4/16

KV 4/65

KV 4/170

KV 4/185

KV 4/211

Secondary sources

Andrew, Christopher, *Defend the Realm. The Authorized History of MI5*, New York, Alfred A. Knopf, 2009.

Batey, Mavis, "Dilly Knox - A Reminiscence of this Pioneer Enigma Cryptanalyst", *Cryptologia*, vol. 32, no. 2, 2008.

Burkett, Randy, "An Alternative Framework for Agent Recruitment: From MICE to RASCLS", *Studies in Intelligence*, vol. 57, no. 1, pp.7-17, 2013.

Churchill, Winston, *War Speeches*, Paris, Tallandier, 2009.

Erskine, Ralph, Smith, Michael (eds), *Action This Day*, London, Bantam Press, 2001.

Farago, Ladislas, *The Game of the Foxes: the Untold Story of German Espionage in the United States and in Great Britain During World War II*, New York, D. McKay, 1971.

Hayward, James, *Double Agent Snow. The True Story of Arthur Owens, Hitler's Chief Spy in England*, London, Simon and Schuster, 2013.

Hinsley, F.H., Simkins, C.A.G., *British intelligence in the Second World War. Vol.4, Security and counter-intelligence*, London, H.M.S.O, 1990.

Holt, Thaddeus, *The Deceivers. Allied Military Deception in the Second World War*, New York, Scribner, 2004.

Kahn, David, *Hitler's Spies*, London, Hodder & Stoughton, 1978.

Macintyre, Ben, *Agent Zigzag, The True Wartime Story of Eddie Chapman. Lover, Traitor, Hero, Spy*, London, Bloomsbury, 2007.

Macintyre, Ben, *Double Cross. The True Story of the D-Day Spies*, London, Bloomsbury, 2012.

Masterman, John Cecil, *The Double Cross System in the War of 1939 to 1945*, New Haven and London, Yale University Press, 1972.

Peis, Günter, Wighton, Charles Wighton, *Hitler's Spies and Saboteurs, Based on the German Secret Service War Diary of General Lahousen*, New York, Holt, 1958.

Stephens, Robin, *Camp 020: MI5 and the Nazi Spies. The Official History of MI5's Wartime Interrogation Centre*, Richmond, PRO, 2000.

West, Nigel, Roberts, Madoc, *Snow. The Double Life of a World War II Spy*, London, Biteback, 2011.

West, Nigel, *MI5: British Security Service Operations, 1909-45*, London, Bodley Head, 1981.

Global Security and Intelligence Studies • Volume 8, Number 1 • Summer 2023

Competitive, Cooperative, and Confounding: Japan and South Korean Relations Through International Relations Theories

Is Getting Past Their Past Possible?

Ken Sandler

U.S. Naval War College
kmsandler154@gmail.com

Abstract

Japan and South Korean relations have consistently confounded traditional International Relations (IR) scholars. Realism assumes that countries with common allies and enemies will be well disposed toward each other. Japan and South Korea's shared security threats should indicate strong bilateral relations. From a liberalism perspective, both are democracies with competitive elections that protect individual rights, have similar mutually appealing cultures, and have interconnected economies. Although both countries cultivated shared strategic goals and mutual friendship at times, antagonism characterizes their bilateral relationship; until recently, considered to be at its lowest point since normalization in 1965.[1] On the other hand, constructivism maintains that national interests and policy are primarily products of identity, rooted in historical memory, ideas, and culture. This comprises the essence of national interests and the purpose of national policies.[2] Emotion-laden collective memories in both countries repeatedly derail temporary improvements.[3] According to South Korea, Japanese atonement for past colonization and war crimes is hollow. At the same time, Japan wonders why the Republic of Korea (ROK) ignores decades of genuine contrition.[4]

---

1   Deacon, Chris. "Reproducing the 'history problem': memory, identity and the Japan-South Korea trade dispute" The Pacific Review, March 10, 2021, 1-3. https://www.tandfonline.com/doi/full/10.1080/09512748.2021.1897652

2   Chaesung, Chun. "A Theoretical Explanation of the Evolving Northeast Asian Architecture: The "Incompleteness" of Sovereignty" The ASAN Forum, January-February 2022, 6. https://theasanforum.org/a-theoretical-explanation-of-the-evolving-northeast-asian-architecture-the-incompleteness-of-sovereignty/

3   Deacon, 3.

4   Chaesung, 6; Cha, Victor. "Entrapment, and Neoclassical Realism in Asia: The United States, Japan, and Korea" International Studies Quarterly, Jun., 2000, Vol. 44, No. 2, 261. https://www.jstor.org/stable/3013998

                                                    doi: 10.18278/gsis.8.1.7

Keywords: Japan, South Korea, international relations, cultural conflict

Competitivo, cooperativo y confuso: Las relaciones entre Japón y Corea del Sur a través de las teorías de las relaciones internacionales

¿Es posible superar su pasado?

RESUMEN

Las relaciones entre Japón y Corea del Sur han confundido constantemente a los estudiosos tradicionales de las Relaciones Internacionales (RI). El realismo supone que los países con aliados y enemigos comunes estarán bien dispuestos entre sí. Las amenazas a la seguridad compartidas por Japón y Corea del Sur deberían indicar relaciones bilaterales sólidas. Desde una perspectiva liberalista, ambas son democracias con elecciones competitivas que protegen los derechos individuales, tienen culturas similares mutuamente atractivas y economías interconectadas. Aunque ambos países cultivaron objetivos estratégicos compartidos y en ocasiones amistad mutua, el antagonismo caracteriza su relación bilateral; hasta hace poco, se consideraba que estaba en su punto más bajo desde la normalización en 1965. Por otro lado, el constructivismo sostiene que los intereses y las políticas nacionales son principalmente productos de la identidad, arraigados en la memoria histórica, las ideas y la cultura. Esto comprende la esencia de los intereses nacionales y el propósito de las políticas nacionales. Los recuerdos colectivos cargados de emociones en ambos países descarrilan repetidamente las mejoras temporales. Según Corea del Sur, la expiación japonesa por la colonización y los crímenes de guerra del pasado es hueca. Al mismo tiempo, Japón se pregunta por qué la República de Corea ignora décadas de genuina contrición.

Palabras clave: Japón, Corea del Sur, relaciones internacionales, conflicto cultural

竞争、合作与困惑
国际关系理论之日韩关系
摒弃前嫌是可能的吗？

摘要

日韩关系一直让传统国际关系(IR)学者感到困惑。现实主义假设，拥有共同盟友和敌人的国家会善待彼此。日本和韩国的共同安全威胁应表明牢固的双边关系。从自由主义的角度来看，两国都是实行竞争性选举的民主国家，保护个人权利、具有类似的相互吸引的文化，并且经济相互关联。尽管两国有时建立了共同的战略目标和友谊，但两国关系具有对立特征；直到最近，日韩关系被认为处于1965年两国关系正常化以来的最低点。另一方面，建构主义认为，国家利益和政策主要是身份的产物，植根于历史记忆、思想和文化。这包括国家利益的本质和国家政策目的。充满情感的集体记忆使两国一再破坏暂时改善的关系。韩国认为，日本对过去殖民和战争罪行的赎罪是空洞的。与此同时，日本想了解为何韩国忽视其数十年的真诚忏悔。

关键词：日本，韩国，国际关系，文化冲突

This essay explores the explanatory power of alternate IR theories, including incomplete sovereignty, poststructuralism, quasi-alliance, and accommodation. Each provides different pieces of the puzzle, with the common theme being the predominant role of historical memory and identity, which creates and sustains distrust. Of these, poststructuralism, a variation of constructivism, most convincingly explains how these ideas establish and maintain a vicious cycle. The dominant role of domestic politics plays the spoiler, with the alignment of Japanese and South Korean leaders and the populations they serve as the most elusive elements of moving beyond the past. Constructivism and its post-structural relative may provide the most compelling explanation for this relationship, while providing hope for changing this cycle into a virtuous one.[5] Robert Putnam's two-level games framework, which examines the bidirectional influence of national and international poli-

---

5   Wendt, Alexander "Anarchy is what States Make of it: The Social Construction of Power Politics," *International Organization* Vol. 46, No. 2 (Spring, 1992), 394-395. http://www.jstor.org/stable/270 6858.

tics on international relations, captures the daunting challenges of making lasting changes.[6]

Incomplete Sovereignty

Chun Chaesung argues that realism, liberalism, and constructivism neglect the impact of "incomplete sovereignty." In this theory, traditional Sinocentric-led East Asian imperialism, along with subsequent Western and Asian imperial practices, created legacies that make it difficult for China, South Korea, and Japan to reconcile their past with modern sovereignty concepts.[7] However, political identity is too simple an explanation. Solving the regional security dilemma requires common aspirations of peaceful coexistence.[8]

Chaesung divides the East Asian order into four periods, in which China, South Korea, and Japan did not gain complete sovereignty. The first was the Chinese hegemonic imperial period before the Opium Wars. At the end of this phase, the West exported the modern state concept to the region. The second was a transitional era from 1840 to 1951. During this time, supposedly Western liberal democratic powers assumed imperial rule to varying degrees in the region. The imprint this left led to Japanese emulation and colonization of Korea and portions of China until World War II ended. The San Francisco Treaty of 1951 attempted to settle regional territorial issues with 48 countries represented. However, neither the two Koreas nor the two Chinas participated, and the Soviet Union did not agree to the outcome. As a result, the treaty solely achieved Japan's bilateral alliance with the US. The author asserts that Japan supported this treaty as its best chance to attain full sovereignty under the circumstances. The third was the Cold War until 1991. From then to the present, the fourth period consists of US-China strategic competition assuming a central role.[9]

East Asia's power transition is intertwined with each country's vision of complete sovereignty. The San Francisco Treaty's shortfalls generated an environment of continuous insecurity where each country sought to gain recognition through various partners, allies, and unilateral actions.[10] According to this theory, South Korea's fears about Japan will only be alleviated when it properly apologizes for past behavior, and as long as Japan's military power is constrained through constitutional provisions. Japan is concerned that a unified Korea could attempt to become a great power at its own expense. Former

---

6   Putnam, Robert D. "Diplomacy and Domestic Politics: The Logic of Two-Level Games" International Organization, Summer, 1988, Vol. 42, No. 3, The MIT Press, 427-460 https://www.jstor.org/stable/2706785.

7   Chaesung, 1.

8   Ibid., 8,6.

9   Ibid., 7.

10   Ibid.

Prime Minister Shinzo Abe's proactive pacifism strategy did not ameliorate South Korean concerns about Japan's imperialist revisionism. Chaesung professes that Japan expressing genuine regret for its past instead of nostalgia would be a major step toward true conciliation and Korean acceptance.[11] He also asserts that both countries are capitalizing on China's rise to achieve their vision of complete sovereignty. South Korea needs China as a constraining force with North Korea. At the same time, China's aggressiveness helps Japanese leaders persuade domestic and international audiences of the need to revise its constitutional military restrictions.[12]

Chun Chaesung makes a solid case that visions of regional sovereignty differ from the modern European definition, making reconciliation difficult. This is due to the stronger roots of imperial systems. Although South Korea and Japan share the common goal of a North Korea without nuclear weapons, their national objectives beyond that diverge, and they are pulled further apart by realist balancing with and against China. While the author accurately diagnoses the need for both countries to reconstruct their vision of peaceful coexistence to satisfy sovereignty concerns, he does not present a visible path to that objective. However, continued North Korean brinksmanship and Chinese assertiveness may lead to a more durable rapprochement. This could result in less-than-full but acceptable bilateral sovereignty. Further, while the timespans are brief compared to the bigger scheme of human existence, they are still broad, especially the transitional phase. This may allow for the patterns he describes to become apparent. However, extended contours of history make it easier to create patterns to fit the theory.

Constructivism/ Poststructuralism

Chris Deacon drew on constructivist/poststructuralist theory through Japan and South Korea's recent trade dispute to analyze how history shapes both nations' identities and foreign policies.[13] He argued that the predominant memories are South Korean "remembering," which portrays Japan as an unrepentant colonizer, and Japanese "forgetting," where South Koreans unjustifiably and unreasonably dwell on the past. These identities require continual reinvigoration to maintain their preeminent influence.[14]

Poststructuralists contend that identity is reliant on dialogue to construct and maintain meaning. This theory subscribes to the importance of group dynamics, in which nationalist self-identity is created and reinforced through opposition to others. Foreign policy practices impact this mutually

---

11   Ibid., 8.

12   Ibid.

13   Deacon, 1.

14   Ibid., 5-6.

constitutive relationship.[15] This is a weak foundation and can evolve, given the proper motivation. In 2019, the trade dispute began when Japan restricted exports of chemicals essential to semiconductor production to South Korea, citing national security concerns.[16] South Korean court decisions, which ruled in favor of compensation for Korean forced labor victims, triggered Tokyo's actions. Japan argued that 1965's normalization treaty settled all prior claims. President Moon's refusal to condemn the rulings and Seoul's subsequent rejection of Japan's request for an arbitration panel led to escalation. South Korea then removed Japan from its preferred trading partner list, and their commerce significantly dropped. This made it impossible to separate both issues.[17]

President Moon continuously accused Japan of creating this controversy. His statements on numerous occasions included 'aggressive attack,' 'deliberate attempt to harm South Korea,' 'deep wounds between both nations,' and 'unfortunate history that the aggressor was reopening.' Moon also sought to capitalize on the dispute for South Korea to become less reliant on others and more competitive in the

semiconductor sector.[18] This media-fueled narrative led to boycotts of Japanese products. Japanese car and beer sales in South Korea dropped tremendously, as did South Korean tourism to Japan. In some instances, actions led to bigotry, with South Korean businesses refusing to admit Japanese customers.[19]

Japan's attempt to sanitize these discourses through international law's supposedly rational and logical prism added fuel to the fire. In this argument, export controls were deliberately decoupled from the forced labor issue and explained as national security concerns, including export to Iran and Syria.[20] South Korea's actions caused Japanese nationalism to reassert itself in a less-apologetic, more proud tradition.[21] Japan portrayed itself as a victim of Korean malicious efforts to rewrite history and focused on its future relationship with South Korea.[22] This episode illustrates foreign policy as a bi-directional extension of domestic politics.[23]

Poststructuralism asserts that there is no objective reality, evidenced by the ROK and Japan's inconsistencies. Although North Korea poses a much more significant threat to South Korea than Japan, the Moon adminis-

---

15   Ibid., 9-10.

16   Ibid., 6.

17   Ibid.

18   Ibid., 11-12.

19   Ibid., 13.

20   Ibid., 14-15.

21   Ibid., 23.

22   Ibid., 16-17.

23   Ibid., 19-20.

tration's overtures to North Korea were bathed in positive language only.[24] In US-Japan relations, remembering the past was an element of reconciliation. After President Obama became the first US President to visit Hiroshima in 2016, Prime Minister Abe followed suit and traveled to Pearl Harbor to commemorate the Japanese attack. The Abe administration never characterized the US as emotional or irrational throughout these visits.[25] The Moon Administration's rhetoric and Abe's response created a vicious loop that has been difficult to break. Although leaders can attempt to steer this narrative onto a more constructive path, this issue is so charged that just acknowledging it can sustain it.[26]

Deacon thoroughly analyzed how this remorseless cycle was created and maintained. Like incomplete sovereignty, though, his optimistic conclusion that change is possible is much more of a sketch lacking detail. At the same time, the dominant conflictual narrative evokes immediate, vivid imagery. However, change is possible, as seen by South Korean President Yoon's outreach to Japan.[27]

Quasi-alliance

Examining 1965 to 1998, Victor Cha tailored alliance theory's balancing to develop his quasi-alliance approach. This theory describes the relationship between two un-allied countries allied to a separate great power patron.[28] Accordingly, Japanese and Korean perceptions of the US's security commitment directly affected their bilateral cooperation much better than the level of external threats.[29]

The quasi-alliance model employs the concepts of abandonment and entrapment to explain the bilateral relationship.[30] Abandonment is fear of the ally leaving the alliance or failure to live up to its commitments when needed. Entrapment is when one country allied to another fears being pulled into a conflict detrimental to its interests. This theory characterizes normal Japan and South Korean relations as friction-prone due to history and different expectations of mutual support. Cha submits that distinguishing between the public's and leadership's hostility is crucial. Emotion and mutual distrust hurt Japanese and Korean interaction at the mass populace level. Abandonment

---

24   Ibid., 20-21.

25   Ibid., 21-22.

26   Ibid., 23.

27   Minegishi, Hiroshi. "Will Japan rebuff overture from South Korea's new leader?" *Nikkei Asia*, April 17, 2022. https://asia.nikkei.com/Spotlight/Comment/Will-Japan-rebuff-overture-from-South-Koreas-new-leader.

28   Cha, 262.

29   Ibid., 261.

30   Ibid., 261.

and entrapment dynamics drive the degree to which this surfaces at the senior leadership level.[31]

Quasi-alliance theory predicts some similar and different behaviors than balance of threat theory. Japan-ROK relations exhibit less contention and greater cooperation when weak American resolve is perceived. However, when the states have different levels of abandonment fears, hostility returns.[32] When US commitments are firm to both countries, as opposed to conventional wisdom, relations will not improve between Japan and South Korea since both are assured of their security. Increased fears of entrapment outweigh additional cohesion.[33] Further, if threats are low but US commitments are weak, the quasi-alliance argument suggests alignment. This is because of the trauma caused by fears of losing security, and the difficulty of internal balancing. Lastly, if concerns about external threats and abandonment are both low, this model predicts bilateral animosity returning.[34]

During the 1969 to 1971 period, the Nixon Doctrine raised acute fears of abandonment by Japan and South Korea. The main thrust of this policy was placing the increased onus on both countries to provide for their security. This led to US force reductions across the region, including the most extensive removal of troops from Korea since the Korean War. It also fully transferred demilitarized zone (DMZ) responsibility to the ROK. South Koreans viewed this as even more alarming due to relatively passive US responses to several DPRK provocations. Both countries' fears of abandonment prompted them to codify the Korea clause and Okinawa base agreement. These agreements effectively committed Japan to South Korea's defense. In addition, Japan provided substantial economic support to South Korea. Further, both sides conducted military exchanges and did not publicize potentially contentious issues.[35]

The remainder of the 1970s witnessed deteriorating followed by improved bilateral ties. Relations worsened from 1972 to 1974 due to different fears of abandonment. For Japan, US-China reconciliation decreased abandonment apprehensions. In contrast, despite 1972's North-South Korea joint communique, communication eventually ceased between them, and abandonment fears remained. South Korea persistently sought US guarantees against further troop reductions and clandestinely pursued a nuclear deterrent.[36] Seoul also vociferously protested the expansion of Japanese-North Korean trade. Even though Japan and South Korea were not allies, Japan sought flexibility and aimed to avoid the percep-

---

31 Ibid., 265.

32 Ibid., 262-263.

33 Ibid., 269.

34 Ibid., 270.

35 Ibid., 275-276.

36 Ibid., 277.

tion of entrapment by South Korea.[37] In 1977, President Carter's efforts to remove all US ground troops from Korea and Congressional involvement via the War Powers Act reinstituted fears of US abandonment in Japan and South Korea.[38] These mutually held fears prompted marked improvement, culminating with unprecedented actions, including the creation of a joint parliamentary council on security affairs, high-level military exchanges, participation in trilateral military exercises, plans for joint military hardware production, and coordinated development of early warning systems.[39]

A similar cycle repeated over the next two decades. The first half of the 1980s once again saw increased friction. Historic summits failed to resolve bilateral disagreements, and instead of alleviating anger, the Emperor's 1984 Showa apology only raised it to new heights. Alleged textbook revisions and disparaging Japanese remarks nearly collapsed economic relations.[40] While the high-level of Soviet and North Korea threat cannot explain this friction, the Reagan Administration's reconfir-

mation of US defense commitments to Asia replaced fears of US abandonment and undermined progress on improved bilateral cooperation.[41] The Cold War's conclusion once again increased fears of US abandonment, leading to considerable cooperation.[42] While the DPRK developed its missile and nuclear capabilities during this period, a conventional invasion similar to the Korean War no longer seemed realistic.[43] Victor Cha advocates the most judicious policy to encourage quasi-allied coordination is US withdrawal over a long period.[44]

Anna Kim claimed the Obama and Trump administrations validated the quasi-alliance theory through periods of cooperation followed by contention.[45] While actions during the Obama Administration seemed to contradict the quasi-alliance theory, she argued that it validated it. During a period of China's ascent and North Korean provocations, the Obama Administration proclaimed its unwavering support for its Asian allies with the 'rebalance' policy.[46] Although firm US resolve would indicate less likelihood of Japan-ROK

---

37   Ibid., 278.

38   Ibid., 279-280.

39   Ibid., 280-281.

40   Ibid., 282.

41   Ibid., 283.

42   Ibid., 284.

43   Ibid., 283.

44   Ibid., 284-285.

45   Kim, Anna. "Understanding Japan–South Korea Cooperation and Friction Through Neoclassical Realist Theory" Jadavpur Journal of International Relations 24(1), 2020, 34. https://journals.sage pub.com/doi/pdf/10.1177/0973598419888279

46   Ibid., 36-37.

cooperation, the US desire for increased burden-sharing and the Budget Control Act's Sequestration mechanism heightened Asian concerns.[47] The pivot to the Pacific failed to curb North Korea's brinkmanship and China's artificial island-building.[48] Although many objected, in 2015, South Korea signed a 'final and irrevocable' agreement with Japan over the comfort women issue.[49]

After sending inconsistent messages, Kim argued that the Trump Administration's commitment to its regional allies validated quasi-alliance theory.[50] Despite Trump's rhetoric concerning burden-sharing and preferences to support South Korean and Japanese nuclear deterrent programs, his administration reassured regional allies once he was elected. Although the US signaled its disdain for multilateral agreements and organizations, the US defense budget drastically rose, regional troop limitations were limited, and the US nuclear arsenal expanded.[51] Pyongyang's nuclear tests and threats gave way to unparalleled summits and warmed relations.[52] This reduced South Korean and Japanese abandonment concerns, worsening bilateral ties be-

tween both countries. Once inaugurated, South Korean President Moon Jae-in's antagonism toward Japan led to their further deterioration.

Victor Cha makes a strong case for the quasi-alliance theory, which contains realist balance of power elements and constructivism's identity based on history. However, Anna Kim's validation of quasi-alliance under the Obama and Trump Administrations is questionable. This theory appears most valid during the bipolar Cold War and post-Cold War unipolar periods, rather than the multipolar period that Kim covers. In his conclusion, Cha acknowledges the same, that his theory may more applicable during the Cold War.[53] Obama's failed rebalance did not lead to strengthened bilateral relationships. Instead of their quarrel subsiding, matters escalated in early 2015 when Japan took a more uncompromising stance on their territorial dispute over Dokdo/Takeshima island. President Park rebuffed Prime Minister Abe's multiple attempts to arrange a summit meeting, sharply different from her warm courtship of Chinese President Xi Jinping.[54] While Kim's conclusion that Japan-ROK rela-

---

47   Ibid., 38-40.

48   Ibid., 38-39; Manyin, Mark E. Daggett, Stephen. Dolven, Ben. Lawrence, Susan V. Martin, Michael F. O'Rourke, Ronald. Vaughn, Bruce. "Pivot to the Pacific? The Obama Administration's "Rebalancing" Toward Asia." Congressional Research Service, March 28, 2012, Introduction. https://sgp.fas.org/crs/natsec/R42448.pdf

49   Ibid., 40-41.

50   Ibid., 41.

51   Ibid., 42-43.

52   Ibid., 43-44.

53   Cooney, Kevin J., and Alex Scarbrough. "Japan and South Korea: Can These Two Nations Work Together?" Asian Affairs35, no. 3, 2008, 182. http://www.jstor.org/stable/30172693.

54   Carpenter, Ted G. "Japan's Growing Quarrel with South Korea: Is China the Main Beneficiary?"

tions worsened during the Trump Administration is accurate, when taken as a whole, US commitment to the region was tenuous, as indicated by the withdrawal from the Trans-Pacific Partnership (TPP) and President Trump skipping the Association of Southeast Asian Nations (ASEAN) senior leaders meeting for two years.[55] Theoretically, this lack of resolve should have improved Japanese-South Korean ties, but that failed to occur. In a reversal of Japan's détente with China in the 1970s, South Korea's strengthened ties with the People's Republic of China (PRC) led to asymmetric relations between Japan, South Korea, and the US. This situation once again fueled a downward trajectory in relations between Japan and South Korea.

Accommodation

David Kang's accommodation theory argued that instead of fearing China's rise and balancing or bandwagoning, South Korea does not regard China as a military threat and sees opportunities to expand its economic interests with the PRC. Therefore, it has adopted a policy of accommodation, which lies between the extremes of balancing and bandwagoning.[56] South Korea has accomplished this without fundamentally changing its security alignment.[57]

South Korea has defied conventional wisdom when it comes to countering the rise of China. Over the last twenty years, South Korea has become closer to China. At the same time, Japanese-South Korean friction has increased to the point of South Korea appearing more concerned about potential Japanese rather than actual Chinese aggression. Although South Korea's security alliance with the US remains secure, there is scant evidence of South Korean balancing.[58] There are pragmatic economic and cultural reasons for this.[59] China is South Korea's largest trading partner, over 25,000 South Korean companies manufacture their products in China, and all major South Korean banks have branch offices there. Culturally, South Korea has seen a 44 percent increase in the number of Chinese language schools. Tourism has greatly expanded, with over four million South Koreans visiting China each year. In addition, more than 54,000 South Koreans studied at universities in China.[60]

CATO Institute, March 5, 2015. https://www.cato.org/commentary/japans-growing-quarrel-south-korea-china-main-beneficiary.

55 Al Jazeera, "Trump snubs meeting with ASEAN leaders in Bangkok" October 30, 2019. https://www.aljazeera.com/news/2019/10/30/trump-snubs-meeting-with-asean-leaders-in-bangkok

56 Kang, David. "Between Balancing and Bandwagoning: South Korea's Response to China" Journal of East Asian Studies 9 (2009), 4. https://www.jstor.org/stable/23418681

57 Ibid., 1.

58 Ibid., 2.

59 Ibid., 4.

60 Ibid., 12-13.

According to Kang, South Korea's foreign policy steadily expanded its relations with China while maintaining its freedom of action.[61] South Korea and China had almost indistinguishable foreign policy positions on various issues, including approaches to North Korea and Japan. South Korea's 2004 National Security Strategy called the China-South Korea relationship a "comprehensive cooperative partnership" and advocated for increased military exchanges.[62] For a period, it seemed that Seoul's foreign policy was shifting closer toward China.[63] While South Korea has consistently maintained its freedom of action, since Kang's 2009 article, China's aggressiveness and North Korea's unpredictability may be shifting South Korea away from the PRC and once again toward Japan.[64]

David Kang makes a strong realist argument why South Korea is drifting into China's orbit due to its interests. He explains that labels for this type of strategy include engagement, accommodation, hiding, and hedging. The most crucial distinction between them is more or less fear of an adversary. Hedging infers a degree of skepti-cism and suspicion about another state, which South Korea does not appear to have.[65] However, another definition of hedging outside of David Kang's describes it as "…leaving strategic options open without balancing or bandwagoning." South Korea's actions meet this more flexible definition.[66]

Analysis

All theories presented include elements of realism and constructivism to different degrees. This includes a consistent theme of historical memory, national identity, and domestic politics. Post-structuralism accurately explains how discourse perpetuates a cycle of hostility while also offering optimism about how changing narratives can impact dynamics. Since foreign policy extends domestic politics, collective memory will undoubtedly play a role. The degree of influence identity has on bilateral relations depends on South Korean and Japanese elites' ability to shape, channel, or resist public sentiment in support of national interests. To paraphrase Thucydides, states are motivated by fear, honor, and

---

61   Ibid., 13-14.

62   Ibid., 9-10.

63   Ibid., 14, 18-19.

64   Randal, Julian. "Will security concerns over China, North Korea push Japan, South Korea to resolve wartime labour dispute?" South China Morning Post, January 17, 2023. https://www.scmp.com/week-asia/politics/article/3206941/will-security-concerns-over-china-north-korea-push-japan-south-korea-resolve-wartime-labour-dispute

65   Ibid., 7.

66   Koga, Kei. "The Concept of "Hedging" Revisited: The Case of Japan's Foreign Policy Strategy in East Asia's Power Shift International Studies Review, 2018, 634. https://academic.oup.com/isr/article/20/4/633/4781685.

interest.[67] In democratic Japan and South Korea, interests include these and run the gamut from state survival to economic benefits and reelection, just to name a few. South Korean animosity toward Japan is always visible or simmering just below the surface, and South Korean honor creates demands that conflict with Japan's honor and interests. South Korean leadership's ability to mend fences with Japan usually went against domestic opposition, even when the threat level was high and there were tangible economic benefits.

One irony is that the 1965 and 2015 agreements between both countries were the products of two President Parks: father and daughter. South Korean President Park Chung-Hee had a more positive perspective concerning the Japanese than his predecessor. During Japan's occupation, he grew up speaking fluent Japanese, was one of few Koreans to attend the Japanese military academy in Tokyo, and fought for Japan during World War II.[68] From his perspective, reconciliation was the best means of developing South Korea's economy. His authoritarian rule was needed to stifle opposition, who accused Park of intriguing with the US to reestablish Japanese domination.[69] The 1965 treaty saw huge protests in Japan and South Korea.[70] These never truly settled the core issues, and the people's passion for this makes any enduring resolution unlikely in the near term. Under President Park Geun-hye, South Korea's first female president, in 2015 Japan and South Korea appeared to finally settle the comfort women issue by Japan's compensation to victims and apologizing. Yet, domestic opposition in both countries led to South Korea scuttling the deal in 2018. Subsequent South Korean court rulings calling into question the 1965 treaty and the longstanding Dokdo/Takeshima islands dispute only added to tensions.[71] President Park later earned her own notoriety as the first ROK leader to be impeached.[72]

However, fear of North Korean or Chinese aggression could bring Japan and South Korea once again closer. Anti-Chinese attitudes have increased

---

67   Cohen, Matthew F. "Strategy's Triumvirate: Thucydides, Sun Tzu, and Clausewitz" On Strategy: A Primer, Army War College, 2020, 28. https://www.armyupress.army.mil/Portals/7/combat-stud ies-institute/csi-books/on-strategy-a-primer.pdf

68   Ibid., 128; Brazinsky, Gregg. "Biographies: Park Chung Hee" Wilson Center Digital Archive. https:// digitalarchive.wilsoncenter.org/resource/modern-korean-history-portal/park-chung-hee

69   Cha, Victor. "Bridging the Gap: The Strategic Context of the 1965 Korea—Japan Normalization Treaty" Korean Studies , 1996, Vol. 20: 129,142.

70   WSJ Staff. "WSJ Archive: 50th Anniversary of Normalization of Japan-South Korea Ties" Wall Street Journal Japan Real-time Blog. June 18, 2015. https://www.wsj.com/articles/BL-JRTB-20202

71   Roehrig, Terence. "The Rough State of Japan–South Korea Relations: Friction and Disputes in the Maritime Domain" Maritime Awareness Project, January 15, 2021: 1-2. https://www.nbr.org/ wp-content/uploads/pdfs/publications/analysis_roehrig_011521.pdf

72   Britannica, "Park Geun-Hye, president of South Korea" https://www.britannica.com/biography/ Park-Geun-Hye

in South Korea. A Pew poll found that South Koreans increasingly have a negative perception of China, rising from 31 percent in 2002 to 75 percent in 2020. This is the result of the PRC's wide-ranging economic sanctions from the ROK's 2016 decision to install the US-sponsored Terminal High Altitude Area Defense (THAAD) missile defense system. Further, a rise in violence against Koreans in China during this period, perceived to be driven by China's state media, added to South Koreans' dissatisfaction with China.[73]

This brings the discussion back to the merits of the quasi-alliance theory. Quasi-alliance qualifies realism's balance of threat predictions with fears of patron abandonment and entrapment. However, its predictions suffer in a multipolar world. Further, its recommendation to withdraw US troops is more challenging within the geopolitics of US-China strategic competition. South Korea and Japan view their relationship with China differently, adding another dimension of complexity to bilateral cooperation.[74] Realism's hedging offers some explanatory and predictive power.[75] Building on the Hobbesian tradition, in which each state acts only in its self-interest, realism's fundamental belief is that international relations are a struggle for power among states.[76] Japan and South Korea share some common interests, but those that pull them apart are greater. Therefore, hedging, an insurance policy against opportunism, will always be present. There was a time when Japan was the nation reaching out to China at South Korea's expense. That role is now reversed. However natural hedging is, it does not fully answer why these two states are unable to sustain cooperation, or which way they will lean. Domestic politics, which sits at the nexus of history and identity, provides the strongest clues.

Domestic Politics

Between the 1965 and 2015 agreements, efforts to repair relations bore some fruit while also sowing the seeds for domestic backlash. Japan's 1993 Kono Statement acknowledged the coercive recruitment of women to serve in comfort stations, which the Japanese military established and managed.[77] Japanese Prime Ministers in the

---

73   Kim, Dongwoo. "The Politics of South Korea's 'China Threat'" The Diplomat, April 05, 2021. https://thediplomat.com/2021/04/the-politics-of-south-koreas-china-threat/

74   Hughes, Christopher. "Japan's Grand Strategic Shift: From the Yoshida Doctrine to an Abe Doctrine?" National Bureau of Asian Research, 2017, 76.

75   Walt, Stephen M. "International Relations: One World, Many Theories." Foreign Policy, Spring 1998, No 110, 43.

76   Snyder, Jack. "One World, Rival Theories" Foreign Policy, Nov-Dec 2004, No 145. P55; Bull, Hedley. The Anarchical Society: a Study of Order in World Politics. New York: Columbia University Press 2002, 22-23; Walt, 36.

77   Sterngold, James. "Japan Admits Army Forced Women Into War Brothels" The New York Times, August 5, 1993: 2. https://www.nytimes.com/1993/08/05/world/japan-admits-army-forced-women-into-war-brothels.html

mid-to-late 1990s regularly issued apologies, and relations reached a brief peak in 1998, with President Kim Dae-Jung inviting Emperor Akihito to visit Korea. An increase in trade and tourism between the two countries followed. Relations once again soured in 2001 after revelations of Japanese schoolbooks describing colonization as necessary for regional security, with any mentions of comfort women removed. Continued controversial visits by Japanese leaders to Yasakuni Shrine embittered many South Koreans.[78]

For each step forward, domestic blowback reversed progress.[79] Transnational solidarity between civic group efforts in Japan and South Korea, working through international organizations like the United Nations (UN) and the US Congress helped publicize comfort women abuses and had some limited impacts. A 1993 survey found that 51 percent of Japanese supported official compensation to surviving comfort women for their forceful conscription. The same poll found that 10.9 percent of the public thought the 1965 Treaty resolved the issue.[80]

In 2007, US House of Representatives Resolution 121 demanded that the Japanese government acknowledge and apologize for war crimes. Japanese conservative groups and media criticized the resolution as unjustified. After Prime Minister Shinzo Abe and his cabinet's inauguration, the issue assumed more prominence.[81] External pressure hardened Japanese sentiment against lasting policy changes to satisfy South Korean public demand. The tide turned during the conservative Shinzo Abe's second cabinet in 2012.[82] Japanese public opinion seemed to solidify against external pressure and the Kono statement after a comfort women statue was erected in Glendale, California in 2013.[83] By 2014, the same question brought about a marked difference. By then, 63 percent of respondents thought the 1965 Treaty adequately resolved the issue and additional compensation was unnecessary. To reinforce this significant change, 42 percent of Japanese desired the Kono Statement's revision.[84] Failure to institutionalize the 2015 agreement in both countries demonstrates the frailty of intergovernmental agreements without solid domestic underpinning.[85]

With a conservative South Korean President and a moderate Japanese

---

78 Japan-guide, "Yasukini Shrine" March 23, 2008. https://www.japan-guide.com/e/e2321.html

79 Kim, Ji Young and Sohn, Jeyong. "Settlement Without Consensus: International Pressure, Domestic Backlash, and the Comfort Women Issue in Japan" Pacific Affairs, Vol. 90, No. 1, March, 2017: 83-84. https://www.jstor.org/stable/44876137

80 Ibid., 77, 79, 80-81.

81 Ibid., 90-92.

82 Ibid., 77, 79-80.

83 Ibid., 94.

84 Ibid., 80-81.

85 Ibid., 98, 81.

Prime Minister, there may be hope to address this contentious issue. In May 2022, the Japanese and South Korean Foreign Ministers met and concurred with the need to improve cooperation. South Korean President Yoon indicated his desire for a summit with Japanese Prime Minister Kishida.[86] This came after continuous US efforts to improve trilateral ties for security and economic reasons, focused on countering China.[87]

Daniel Sneider recommended a new comprehensive agreement between both countries, which would be fashioned on Germany's Foundation for Remembrance, Responsibility and the Future to compensate former forced laborers. It would succeed the now defunct Reconciliation and Healing Foundation. Unlike the 2015 agreement, this would involve the Japanese government and corporations. As Sneider admits, the domestic barriers to such an arrangement are discouraging in both countries. Victor Cha submitted the importance of distinguishing between the public's and leadership's hostility.

Even with senior leaders in agreement, at the mass populace level, emotion and mutual distrust hurt Japanese and Korean interaction.[88]

Two-Level Games

Following the trilateral summit in late June 2022 between Japan, South Korea, and the US, Japan's Prime Minister Fumio Kishida and South Korea's President Yoon Suk-yeol met on September 21, 2022 at the UN General Assembly, where they expressed their shared interests and desire to loosen tense bilateral relations. This subdued informal meeting was deliberately devoid of the usual fanfare to avoid domestic backlash.[89] However, the Yoon administration faces several constraints on its ability to negotiate with Japan including the South Korean Supreme Court's ruling on wartime labor, low public approval ratings close to 30 percent, and an opposition Democratic Party eager to attack perceived concessions.[90] Likewise,

---

86 Park, Jumin. "Japan Says S.Korea Wants to Work on Improving Ties" Reuters, May 10, 2022. https://www.usnews.com/news/world/articles/2022-05-10/s-korea-wants-to-work-on-improving-bilateral-ties-japanese-foreign-minister

87 Toyoura, Junichi. "South Korea President Yoon begins pursuit of improved relations with Japan" Asia News Network, June 4, 2022. https://asianews.network/south-korea-president-yoon-begins-pursuit-of-improved-relations-with-japan/; Harris, Tobias. "Fumio Kishida's Principles Are About to Be Put to the Test" October 4, 2021. Foreign Policy, https://foreignpolicy.com/2021/10/04/fumio-kishida-new-japanese-prime-minister-ldp/

88 Sneider, Daniel. "Cutting the Gordian knot in South Korea–Japan relations" EastAsia Forum, April 4, 2022. https://www.eastasiaforum.org/2022/04/04/cutting-the-gordian-knot-in-south-korea-japan-relations/

89 Minegishi, Hiroshi. "Three 'cushions' help Japan, South Korea leaders hold first talks" Nikkei Asia, October 3, 2022. https://asia.nikkei.com/Spotlight/Comment/Three-cushions-help-Japan-South-Korea-leaders-hold-first-talks

90 Shin, Mitch. "South Korea's Yoon Suk-yeol Finally Meets Japan's Prime Minister" The Diplomat, September 22, 2022. https://thediplomat.com/2022/09/south-koreas-yoon-suk-yeol-final-

Prime Minister Kishida's options have been limited. He is pressured by his own Liberal Democratic Party (LDP) to withhold a summit until progress on resolving the wartime laborers issue.[91] In a mid-September 2022 poll, Kishida's Cabinet approval rating dropped to 40.2 percent, a new low from 63 percent in its sweeping victory in August.[92] Further, in early October, Japanese opposition parties including the main Constitutional Democratic Party, increased their criticism of Kishida on separate issues. These setbacks cumulatively hamper Kishida's agenda, including delicate efforts to improve ties with South Korea.[93]

Unlike quasi-alliance's prediction that strong US commitment to both countries would exacerbate Japan-ROK relations, today's multipolar world may be demonstrating the critical role the US can play in reconciling both nations. Tokyo and Seoul developed a "three-cushion" approach to mitigate some of the domestic obstacles to strengthened relations. This consists of US involvement, discussions outside of both countries, and use of an international framework. At the trilateral meeting in Hawaii, both countries national security advisors discussed the importance of sincerely resolving long-standing and more recent disputes. Confidence building steps include South Korea's Prime Minister visiting Japan for the first time in three years to attend Shinzo Abe's state funeral. Despite the LDP's opposition, Japan responded by inviting South Korea to its fleet review.[94]

Minefields abound as this process unfolds, and there is always the chance that the undercurrents of opposition to any arrangement will sweep away attempts to build bridges. The Biden Administration's Inflation Reduction Act, which offers tax credits for purchases of US-assembled electric vehicles, adversely impacts South Korean car manufacturers. South Korea claims this discriminatory practice violates free trade principles. It is too early to tell if this will harm US efforts to foster Japan-ROK reconciliation although it has the potential to harden one of

---

ly-meets-japans-prime-minister/; Minegishi, Hiroshi. "Three 'cushions' help Japan, South Korea leaders hold first talks" October 3, 2022. https://asia.nikkei.com/Spotlight/Comment/Three-cush ions-help-Japan-South-Korea-leaders-hold-first-talks.

91 Minegishi, Hiroshi. "Three 'cushions' help Japan, South Korea leaders hold first talks" Nikkei Asia, October 3, 2022. https://asia.nikkei.com/Spotlight/Comment/Three-cushions-help-Japan-South-Korea-leaders-hold-first-talks

92 Staff Report, "Kishida Cabinet support rate continues to plummet, survey shows" The Japan Times, September 18, 2022. https://www.japantimes.co.jp/news/2022/09/18/national/politics-diplomacy/fumio-kishida-cabinet-survey-lowest/

93 Kyodo News, "Japan's opposition steps up offensive against beleaguered PM Kishida" October 5, 2022. https://english.kyodonews.net/news/2022/10/6647c066bb8e-kishida-defends-minister-with-suspicious-ties-with-unification-church.html

94 Minegishi, Hiroshi. "Three 'cushions' help Japan, South Korea leaders hold first talks" Nikkei Asia, October 3, 2022. https://asia.nikkei.com/Spotlight/Comment/Three-cushions-help-Japan-South-Korea-leaders-hold-first-talks

the three "cushions" critical to this approach.[95] This scenario could produce asymmetric relations between all three nations and would be another test of quasi-alliance's prediction that this would once again degrade Japan-ROK relations.

Constructivism argues that historical identity and norms are learned processes that can be unlearned.[96] Within a backdrop of recent North Korean missile tests and China's assertiveness, persistent US diplomatic efforts, and a framework which acknowledges the importance of public sentiment on both sides, there may be an opportunity to achieve lasting resolution.[97] Successful Track II academic cooperation and cultural exchanges are one means of providing upwards pressure to change the narrative.[98] Another is the significant improvement in public attitude surveys. Between 2021 and 2022,

the number of South Korean respondents who had a negative opinion of Japan dropped 10.4 percent, while Japanese responded similarly, dropping 8.5 percent. During the same time, the number of South Koreans who had a positive opinion of Japan climbed 20.5 percent, while Japanese opinion toward South Korea improved 5 percent.[99]

Constructivism also posits that nations act differently towards enemies than friends.[100] With compromise the root of peaceful solutions, perhaps a happy medium in the face of shared external threats that outweighs mutual animosity will incrementally build a virtuous cycle that eventually satisfies the demands of both publics. That may come to fruition as both countries take positive steps toward resolving the wartime labor dispute.[101] Although many have said that countries do not have friends, only interests, changing the un-

95 Gallo, William. "'Worse Than MAGA' – South Korea Erupts Over Biden's Trade Policy" Voice of America, September 14, 2022. https://www.voanews.com/a/worse-than-maga-south-korea-erupts-over-biden-s-trade-policy-/6747016.html

96 Wendt, Alexander "Anarchy is what States Make of it: The Social Construction of Power Politics," International Organization Vol. 46, No. 2 (Spring, 1992), 396-397. http://www.jstor.org/stable/2706858.

97 Dominguez, Gabriel. "North Korean provocations boosting U.S., Japan and South Korea security cooperation" The Japan Times, October 6, 2022. https://www.japantimes.co.jp/news/2022/10/06/national/politics-diplomacy/japan-us-south-korea-cooperation-pyongyang/

98 Cooney, Kevin J., and Alex Scarbrough. "Japan and South Korea: Can These Two Nations Work Together?" Asian Affairs35, no. 3, 2008, 177. http://www.jstor.org/stable/30172693.

99 Genron NPO, "Will improved public sentiment in Japan and South Korea lead to better relations between the two countries?" September 12, 2022. https://www.genron-npo.net/en/opinion_polls/archives/5605.html

100 Wendt, Alexander "Anarchy is what States Make of it: The Social Construction of Power Politics," International Organization Vol. 46, No. 2 (Spring, 1992), 396-397. http://www.jstor.org/stable/2706858.

101 Kyodo News, "Japan to Uphold Apology if South Korea settles wartime labor issue" January 28, 2023. https://english.kyodonews.net/news/2023/01/b344d7acf82c-japan-to-uphold-apology-if-s-korea-settles-wartime-labor-issue.html

derlying narrative is a precursor to moving beyond the past.[102]

LtCol **Ken Sandler**, U.S. Marine Corps (retired), served as a military professor at the U.S. Naval War College from 2018-2023. The views presented are those of the author and do not necessarily reflect the policy or position of the U.S. Marine Corps, the U.S. Navy, the Naval War College, or the Department of Defense. He holds a Masters of Science in Information Systems Technology from George Washington University and Masters of Strategic Studies from the U.S. Army War College. His certifications include Project Management Professional (PMP) and Certified Information Systems Security Professional (CISSP). For most of his career, he supervised organizational telecommunications networks and cybersecurity postures. He is currently pursuing a PhD in International Relations.

References

Al Jazeera, "Trump snubs meeting with ASEAN leaders in Bangkok" October 30, 2019. https://www.aljazeera.com/news/2019/10/30/trump-snubs-meeting-with-asean-leaders-in-bangkok

Babones, Salvatore. "Trump Has Alienated Allies—but Has Them Acting in America's Interest (and Their Own)" Foreign Policy, August 6, 2020. https://foreignpolicy.com/2020/08/06/trump-foreign-policy-accomplishments

Brazinsky, Gregg. "Biographies: Park Chung Hee" Wilson Center Digital Archive. https://digitalarchive.wilsoncenter.org/resource/modern-korean-history-portal/park-chung-hee

Britannica, "Park Geun-Hye, president of South Korea" https://www.britannica.com/biography/Park-Geun-Hye

Carpenter, Ted G. "Japan's Growing Quarrel with South Korea: Is China the Main Beneficiary?" CATO Institute, March 5, 2015. https://www.cato.org/commentary/japans-growing-quarrel-south-korea-china-main-beneficiary.

Cha, Victor. "Entrapment, and Neoclassical Realism in Asia: The United States, Japan, and Korea" International Studies Quarterly, Jun., 2000, Vol. 44, No. 2, 261. https://www.jstor.org/stable/3013998

---

102 Babones, Salvatore. "Trump Has Alienated Allies—but Has Them Acting in America's Interest (and Their Own)" Foreign Policy, August 6, 2020. https://foreignpolicy.com/2020/08/06/trump-foreign-policy-accomplishments/

Cha, Victor. "Bridging the Gap: The Strategic Context of the 1965 Korea—Japan Normalization Treaty" Korean Studies , 1996, Vol. 20: 129,142.

Cohen, Matthew F. "Strategy's Triumvirate: Thucydides, Sun Tzu, and Clausewitz" On Strategy: A Primer, Army War College, 2020, 28. https://www.armyupress. army.mil/Portals/7/combat-studies-institute/csi-books/on-strategy-a-primer.pdf

Chaesung, Chun. "A Theoretical Explanation of the Evolving Northeast Asian Architecture: The "Incompleteness" of Sovereignty" The ASAN Forum, January-February 2022, 6. https://theasanforum.org/a-theoretical-explanation-of-the-evolvin g-northeast-asian-architecture-the-incompleteness-of-sovereignty/

Cooney, Kevin J., and Alex Scarbrough. "Japan and South Korea: Can These Two Nations Work Together?" Asian Affairs35, no. 3, 2008, 182. http://www.jstor.org/ stable/30172693.

Deacon, Chris. "Reproducing the 'history problem': memory, identity and the Japan-South Korea trade dispute" The Pacific Review, March 10, 2021, 1-3. https:// www.tandfonline.com/doi/full/10.1080/09512748.2021.1897652

Dominguez, Gabriel. "North Korean provocations boosting U.S., Japan and South Korea security cooperation" The Japan Times, October 6, 2022. https://www.japan times.co.jp/news/2022/10/06/national/politics-diplomacy/japan-us-south-ko rea-cooperation-pyongyang/

Gallo, William. "'Worse Than MAGA' – South Korea Erupts Over Biden's Trade Policy" Voice of America, September 14, 2022. https://www.voanews.com/a/worse- than-maga-south-korea-erupts-over-biden-s-trade-policy-/6747016.html

Genron NPO, "Will improved public sentiment in Japan and South Korea lead to better relations between the two countries?" September 12, 2022. https://www. genron-npo.net/en/opinion_polls/archives/5605.html

Hughes, Christopher. "Japan's Grand Strategic Shift: From the Yoshida Doctrine to an Abe Doctrine?" National Bureau of Asian Research, 2017, 76.

Japan-guide, "Yasukini Shrine" March 23, 2008. https://www.japan-guide.com/e/ e2321.html

Kang, David. "Between Balancing and Bandwagoning: South Korea's Response to China" Journal of East Asian Studies 9 (2009), 4. https://www.jstor.org/stabl e/23418681

Kim, Anna. "Understanding Japan–South Korea Cooperation and Friction Through Neoclassical Realist Theory" Jadavpur Journal of International Relations 24(1), 2020, 34. https://journals.sagepub.com/doi/pdf/10.1177/0973598419888279

Kim, Dongwoo. "The Politics of South Korea's 'China Threat'" The Diplomat, April 05, 2021. https://thediplomat.com/2021/04/the-politics-of-south-koreas-china-threat/

Kim, Ji Young and Sohn, Jeyong. "Settlement Without Consensus: International Pressure, Domestic Backlash, and the Comfort Women Issue in Japan" Pacific Affairs, Vol. 90, No. 1, March, 2017: 83-84. https://www.jstor.org/stable/44876137

Koga, Kei. "The Concept of "Hedging" Revisited: The Case of Japan's Foreign Policy Strategy in East Asia's Power Shift International Studies Review, 2018, 634. https://academic.oup.com/isr/article/20/4/633/4781685.

Kyodo News, "Japan to Uphold Apology if South Korea settles wartime labor issue" January 28, 2023. https://english.kyodonews.net/news/2023/01/b344d7acf82c-japan-to-uphold-apology-if-s-korea-settles-wartime-labor-issue.html

Kyodo News, "Japan's opposition steps up offensive against beleaguered PM Kishida" October 5, 2022. https://english.kyodonews.net/news/2022/10/6647c066bb8e-kishida-defends-minister-with-suspicious-ties-with-unification-church.html

Manyin, Mark E. Daggett, Stephen. Dolven, Ben. Lawrence, Susan V. Martin, Michael F. O'Rourke, Ronald. Vaughn, Bruce. "Pivot to the Pacific? The Obama Administration's "Rebalancing" Toward Asia." Congressional Research Service, March 28, 2012, Introduction. https://sgp.fas.org/crs/natsec/R42448.pdf

Minegishi, Hiroshi. "Three 'cushions' help Japan, South Korea leaders hold first talks" Nikkei Asia, October 3, 2022. https://asia.nikkei.com/Spotlight/Comment/Three-cushions-help-Japan-South-Korea-leaders-hold-first-talks

Minegishi, Hiroshi. "Will Japan rebuff overture from South Korea's new leader?" Nikkei Asia, April 17, 2022. https://asia.nikkei.com/Spotlight/Comment/Will-Japan-rebuff-overture-from-South-Korea-s-new-leader.

Park, Jumin. "Japan Says S.Korea Wants to Work on Improving Ties" Reuters, May 10, 2022. https://www.usnews.com/news/world/articles/2022-05-10/s-korea-wants-to-work-on-improving-bilateral-ties-japanese-foreign-minister

Putnam, Robert D. "Diplomacy and Domestic Politics: The Logic of Two-Level Games" International Organization, Summer, 1988, Vol. 42, No. 3, The MIT Press,

427-460 https://www.jstor.org/stable/2706785.

Randal, Julian. "Will security concerns over China, North Korea push Japan, South Korea to resolve wartime labour dispute?" South China Morning Post, January 17, 2023. https://www.scmp.com/week-asia/politics/article/3206941/will-securi ty-concerns-over-china-north-korea-push-japan-south-korea-resolve-wartime-labour-dispute

Roehrig, Terence. "The Rough State of Japan–South Korea Relations: Friction and Disputes in the Maritime Domain" Maritime Awareness Project, January 15, 2021: 1-2. https://www.nbr.org/wp-content/uploads/pdfs/publications/analysis_roehrig _011521.pdf

Shin, Mitch. "South Korea's Yoon Suk-yeol Finally Meets Japan's Prime Minister" The Diplomat, September 22, 2022. https://thediplomat.com/2022/09/south-koreas-yoon-suk-yeol-finally-meets-japans-prime-minister/; Minegishi, Hiroshi. "Three 'cushions' help Japan, South Korea leaders hold first talks" October 3, 2022. https://asia.nikkei.com/Spotlight/Comment/Three-cushions-help-Japan-South-Korea-leaders-hold-first-talks.

Sneider, Daniel. "Cutting the Gordian knot in South Korea–Japan relations" EastAsia Forum, April 4, 2022. https://www.eastasiaforum.org/2022/04/04/cutting-the-gordian-knot-in-south-korea-japan-relations/

Snyder, Jack. "One World, Rival Theories" Foreign Policy, Nov-Dec 2004, No 145. P55; Bull, Hedley. The Anarchical Society: a Study of Order in World Politics. New York: Columbia University Press 2002, 22-23; Walt, 36.

Staff Report, "Kishida Cabinet support rate continues to plummet, survey shows" The Japan Times, September 18, 2022. https://www.japantimes.co.jp/news/2022 /09/18/national/politics-diplomacy/fumio-kishida-cabinet-survey-lowest/

Sterngold, James. "Japan Admits Army Forced Women Into War Brothels" The New York Times, August 5, 1993: 2. https://www.nytimes.com/1993/08/05/world/ japan-admits-army-forced-women-into-war-brothels.html

Toyoura, Junichi. "South Korea President Yoon begins pursuit of improved relations with Japan" Asia News Network, June 4, 2022. https://asianews.network/ south-korea-president-yoon-begins-pursuit-of-improved-relations-with-japan/; Harris, Tobias. "Fumio Kishida's Principles Are About to Be Put to the Test" October 4, 2021. Foreign Policy, https://foreignpolicy.com/2021/10/04/fumio-kishida-new-japanese-prime-minister-ldp/

Walt, Stephen M. "International Relations: One World, Many Theories." Foreign Policy, Spring 1998, No 110, 43.

Wendt, Alexander "Anarchy is what States Make of it: The Social Construction of Power Politics," International Organization Vol. 46, No. 2 (Spring, 1992), 394-395. http://www.jstor.org/stable/2706858.

WSJ Staff. "WSJ Archive: 50th Anniversary of Normalization of Japan-South Korea Ties" Wall Street Journal Japan Real-time Blog. June 18, 2015. https://www.wsj.com/articles/BL-JRTB-20202

Global Security and Intelligence Studies • Volume 8, Number 1 • Summer 2023

Social Media Lies: User's Private Information and the Spread of Misinformation and Disinformation

Christina Roberts, Keith Ludwick, Ph.D.

Abstract

Social media is becoming more and more integrated with everyday life. Individuals around the globe use social media to stay connected with family and friends, conduct business, market products, run political campaigns, and a myriad of other activities. Unfortunately, the integration of social media with daily life is becoming so ubiquitous as to numb users to the possibility that the companies controlling our access to social media might be using the private information gleaned from individual users to help spread misinformation/disinformation (MIDI). This paper asks: How does the use of private information by social media companies lead to the spread of misinformation or disinformation? Using a Grounded Theory approach and examining three instances of social media companies accessing users' private information (Cambridge Analytica, Russian interference in the 2016 U.S. Presidential Election, and the January 6th insurrection at the U.S. Capitol), this paper interprets these incidents to determine the extent of culpability of the company's use of private user information. The findings indicate a strong link between these companies' private, personal information use and the spread of MIDI. The study finishes with policy recommendations and suggestions for future research.

Keywords: social media; disinformation; cognitive warfare; advertising; consent

Mentiras en las redes sociales: información privada del usuario y difusión de información errónea y desinformación

Resumen

Las redes sociales están cada vez más integradas en la vida cotidiana. Personas de todo el mundo utilizan las redes sociales para mantenerse en contacto con familiares y amigos, realizar negocios, comercializar productos, realizar campañas políticas y muchas otras

doi: 10.18278/gsis.8.1.8

actividades. Desafortunadamente, la integración de las redes sociales con la vida diaria se está volviendo tan omnipresente que insensibiliza a los usuarios ante la posibilidad de que las empresas que controlan nuestro acceso a las redes sociales puedan estar utilizando la información privada obtenida de usuarios individuales para ayudar a difundir información errónea/desinformación (MIDI). Este artículo pregunta: ¿Cómo el uso de información privada por parte de las empresas de redes sociales conduce a la difusión de información errónea o desinformación? Utilizando un enfoque de teoría fundamentada y examinando tres casos de empresas de redes sociales que acceden a información privada de los usuarios (Cambridge Analytica, la interferencia rusa en las elecciones presidenciales de EE. UU. de 2016 y la insurrección del 6 de enero en el Capitolio de EE. UU.), este artículo interpreta estos incidentes para determinar el grado de culpabilidad del uso por parte de la empresa de información privada del usuario. Los hallazgos indican un fuerte vínculo entre el uso de información personal y privada de estas empresas y la difusión de MIDI. El estudio finaliza con recomendaciones de políticas y sugerencias para futuras investigaciones.

Palabras clave: redes sociales; desinformación; guerra cognitiva; publicidad; consentir

社媒谎言：用户私人信息与错误及虚假信息的传播

摘要

社交媒体越来越融入日常生活。全球各地，人们使用社交媒体与家人和朋友保持联系、开展业务、营销产品、开展政治活动以及进行无数其他活动。不幸的是，社交媒体与日常生活的融合变得如此普遍，以至于用户对一种可能性变得麻木，即控制我们访问社交媒体的公司可能会利用从个人用户收集的私人信息来帮助传播错误信息/虚假信息(MIDI)。本文的研究问题是：社交媒体公司对私人信息的使用如何导致错误信息或虚假信息的传播？通过使用扎根理论法并研究社交媒体公司获取用户私人信息的三个实例（"剑桥分析"事件、俄罗斯干预2016年美国总统选举、以及1·6美国国会暴乱事件），本文对这些事件进行了诠释，以确定公司使用私人用户信息一事的责任程度。分析结果表明，这些公司对个人信息的使用与MIDI的传播之间存在密切联系。本研究最后提出了政策建议和对未来研究的建议。

关键词：社交媒体，虚假信息，认知战，广告，同意

Introduction

The meteoric rise of social media into the fabric of global society over the past twenty years has been unprecedented. Not since the internet itself has technology become as quickly intertwined with daily life. The benefits of social media are well documented: from connecting friends, organizing advocacy groups, and communicating with victims of disaster, to forums to share information and listing employment opportunities, social media offers a myriad of ways to help society (George, Rovniak, and Kraschnewski 2013; Houston et al. 2015; Sethi 2013). However, there is a darker side: scammers, misinformation/disinformation (called MIDI) campaigns from individuals and nation-states, as well other criminals all take advantage of the unregulated and ubiquitous nature of social media platforms.

Nevertheless, while these nefarious individuals taking advantage of social media represent a threat to the everyday user, a bigger threat is just starting to come to light, that of the social media conglomerates and corporations themselves. March through October of 2021, the U.S. Congress held the first hearings of its kind investigating the potential of social media companies using their respective platforms to gather private information on customers, allow the spread of MIDI, and other practices for which their users might be unaware ("Disinformation Nation: Social Media's Role in Promoting Estremism and Misinformation" 2021). From their inception, most of the companies (e.g., Facebook—now Meta—, Twitter, YouTube) took a "hands-off" approach with the argument that social media were platforms of free speech, and all voices should be heard (Brannon 2018). But that began to change somewhat in the 2010s as social media became platforms for hate speech and radicalization for groups such as white supremacists and Islamic terrorists (Asongu and Orim 2019).

Cries from politicians on both sides of the aisle began to surface that social media companies were supporting these violent organizations with their absence of censorship. Facebook CEO Mark Zuckerberg famously responded that social media's role should be dedicated to the free expression of ideas (Abril 2019). These shouts from politicians and the public were amplified during the COVID pandemic, when Facebook and other platforms were accused of either spreading MIDI about vaccines or squashing alternative views on the efficacy of vaccines or other medicines (Cascini et al. 2022). During 2019-2021 in particular, the role of social media platforms and the spread of (at worst) deliberately fabricated information or (at best) incorrect information innocently passed along started to come more into question.

The issue of social media companies' responsibility to help stop the spread of MIDI has yet to resolve, nor is it likely to be fully addressed in the near term. Whether self-management by the companies will come into play or by government regulation will offer changes remains to be seen; however, the fact that these social media conglomerates play *some* role in spreading MIDI is undeniable. How these companies attract and keep users staring at computer screens and mobile phones is often a closely guarded secret as it relies on finely tuned algorithms that continue to evolve with the sole purpose of keeping users glued to their electronic devices. While it varies somewhat from company to company, the end result is that if users take part in social media platforms, their hosts subject them to advertising specifically targeted to users' needs or wants (Brannon 2018). By feeding users information that incentivizes them to keep viewing similar information, social media companies continue to generate revenue.

The other part of this model is the collection of personal user information. Few people read the ubiquitous "Terms and Agreements" page when downloading a social media app, but buried in the text is the acknowledgment that all activity conducted by a user while using an application is captured and stored (Thompson 2015). While things like names, dates, and photos seem obvious, many of today's users would be shocked to know that social media companies capture minute granular information such as how long a user stays on a particular story or video, the content that they "like" or comment on, the type of cell phone they are using, and time of day when browsing through various stories/postings (Lee and Willett 2021). This private information is captured, analyzed, and often used for targeting advertising, algorithm fine-tuning, or possibly even sold to other companies (Lee and Willett 2021)

Following this line of thinking, it stands to reason that personal user information gathered by social media companies is somehow, at least partially, responsible for the spread of MIDI. This leads to the query addressed in this paper, how does the use of private information by social media companies lead to the spread of misinformation or disinformation? Ultimately, this study hypothesizes that social media companies use of private information has a significant and measurable impact on the spread of disinformation and misinformation due to targeting algorithms and social media companies' financial incentives.

Purpose Statement

This research aims to bring attention to the growing problem of disinformation and misinformation being spread by social media companies use of customers' personal and private information. It is presumed that the spread of MIDI is universally acknowledged to be a negative consequence of the growth of social media over the past decade. What current research has not addressed is how companies are using individuals' private information to help support the spread of MIDI. Investigations into how this occurs, and the ramifications of the use of private information for the spread of "bad" information, will ultimately help academics further understand the far-reaching impact of the use of private information as well as aid policymakers in possible regulatory approaches to help stem the rising tide of a possible increase in spreading MIDI by all types of individuals, whether their motives are criminal or to influence nation-states.

Consider Facebook's role in the spread of antivaccine rhetoric. Researchers from a U.S.-based human rights nonprofit organization ran a two-day test with brand new Facebook accounts to see how quickly Facebook's algorithm sent them down an antivaccine "rabbit hole" (Avaaz 2021). One account searched for the word "vaccine" and liked pages recommended to the account, including antivaccine content, while the other liked a single well-known page that shared antivaccine content (2021). In both cases, the researchers found that Facebook's "you may also like" recommendations included antivaccine content marked with Facebook's warning label for misinformation about COVID-19 or vaccines (2021). During the two-day course of the research, Facebook recommended over 109 antivaccine pages, with a total of 1.4 million followers to the two accounts (2021). This small test demonstrates the concerning pattern Facebook has of promoting disinformation and misinformation to raise the engagement numbers and increase advertising revenue.

As another example involving antivaccine rhetoric, the Center for Countering Digital Hate (CCDH) found that over 65% of the disinformation about COVID vaccines seen on Twitter and Facebook came from what they coined "the disinformation dozen" (CCDH 2021). Previous research found that social media platforms consistent-ly failed to act on 95% of COVID and vaccine disinformation reported to the platforms (2021). This includes Facebook's failure to remove posts sharing content from the Disinformation Dozen, such as YouTube videos. As of July 2021, half of the YouTube channels identified by the CCDH were still available, searchable, and posting more content (De Vynck and Lerman 2021). The failure to act on known disinformation indicates the platforms lack significant incentives to remove such information.

Literature Review

Research into social media and its impact on culture, community, and society is nascent but growing rapidly. Entire journals are now dedicated to investigating social media and its impact on the psychological, sociological, cultural, and political aspects of global society (e.g., see *Social Media + Society, Journal of Digital & Social Media Marketing, and Journal of Social Media and Society)*. This blossoming field of inquiry is currently "finding its way" toward consensus on various issues, including the psychological impact of social media on adolescents (Smith, Leonis, and Anandavalli 2021), the use of social media in intelligence/military operations (Sufi and Alsulami 2022), and business development (Tourani 2022). While these various topics are relevant, because of the broad spectrum of academic discourse available, it is necessary to narrow the literature down accordingly; therefore, the research posed by this paper will focus on social media and the intersection between private information and its use in the spread of MIDI by companies.

The literature surrounding the specific aspects of privacy, social media, and MIDI is broad. However, this study has three themes of interest: personal information privacy, MIDI, and fact-checking within social media. While the academic debate is still ongoing regarding these themes, some thoughts can be generalized.

Personal information/Privacy

As users scroll through Facebook, they can reshare content, create new posts with text or images, react to posts using emojis, "like" or "dislike" postings, leave comments, or send messages to other users (Franz et al. 2019). Each of these actions creates data (2019). Data includes three types: data the user gives voluntarily, purchased data (from a data broker or other organization), and auto-generated data like user analytics (psychological profiles created from what a user interacts with or quizzes they take) (Unger 2020). Unger takes the approach that the absence of personal data privacy protection is the "root cause" of MIDI (Unger 2020, 309). He argues that there is a systematic effort, particularly in political campaigns, to spread MIDI based on the lack of data privacy protection (Unger 2020).

Data privatization may offer different levels of anonymity to varying degrees of success. In Ghazaleh Beigi's Ph.D. dissertation, she concurs to an extent with Unger but draws out a different conundrum, that of privacy vs. utility (Beigi 2020). She used several methods to anonymize a set of social media users' data. Despite anonymizing the structural and textual aspects of a social media data set, the Athd-Improved algorithm reidentified at least 56% of the users in the anonymized dataset (Beigi 2020). This research demonstrates a lack of ability to successfully anonymize to the level needed to maintain privacy and utility. This leads to the question of whether users understand the value of their data or the need for data privacy.

A 2019 survey of 2,416 Americans found that the median consumer would pay only $5 a month for data privacy but would want an $80 purchase price before being willing to share that data (Winegar and Sunstein 2019). The researchers surmised that the high disparity between "willingness to pay" ($5) and "willingness to accept" ($80) illustrates a significant lack of understanding, on the consumers' behalf, about the worth of their data privacy (2019). This disparity constructs the current paradox of consumers valuing their privacy when offered to them but showing a lack of inclination to add privacy to their budgets.

A survey of 530 university respondents by Gary Hunter and Steven Taylor found that the more people value anonymity, the more likely they spend time on social media. Although, respondents who particularly valued seclusion and privacy spent less time on social media (Hunter and Taylor 2019). While this survey was small, it may predict similar patterns within the general population. The next question is whether options exist for those valuing privacy to limit the use of their data by others effectively.

Misinformation and Disinformation

Misinformation is false, erroneous, or incorrect knowledge or evidence; actors spread misinformation due to mistakes, errors, or false beliefs in good faith (Wilczek 2020). Disinformation poses as a more nefarious cousin to misinformation, where malicious actors intentionally spread malicious information to cause harm or achieve objectives (Wilczek 2020; Baptist and Gluck 2021). Disinformation consists of everything from false stories and deceptive advertising to government propaganda websites impersonating more reputable information sources (Goldstein 2021; Fallis 2015). Disinformation is deliberate, whereas misinformation stems from incorrect beliefs or mildly obscured truths (Goldstein 2021; Li and Li 2013). For example, a state-sponsored messaging campaign to promote falsehoods about vaccine harm would constitute disinformation, but in misinformation, a person would share a meme from the disinformation campaign in good faith, fully believing what they read and shared (Barnes 2021; University of Washington 2021).

Fact-Checking

Fact-checking is the effort by an organization to verify the veracity of information from sources. Fact-checkers can be in-house or contracted out and have long been a staple of traditional journalism (Cotter, DeCook, and Kanthawala 2022). In recent years, in part to appease politicians and an ever-growing skeptical public, social media companies and other less well-known information dealers have started to employ fact-checkers to help them establish some boundaries on published information. These fact-checkers can be automated, which comes with its own set of problems.

Cotter et al. state that "...ensuring that the global public are well-informed requires assurance that platform fact-checking functions effectively and in the public's best interest" (Cotter, DeCook, and Kanthawala 2022, 1). Generally, people think of fact-checking as a positive step in the quest for unadulterated information published in social media and other venues, but questions remain as to the standards (which vary from company to company) and intent (Cotter, DeCook, and Kanthawala 2022).

Dubois et al. take a similar approach by demonstrating how the employment of fact-checkers helps eliminate the "echo chamber" effect often seen as the root of the problem of social media (Dubois and Blank 2018). Dubois et al. further state that social media platforms generate trust for users by employing fact-checkers, but until standards are created and met, user doubts will remain (Dubois and Blank 2018).

Gaps

As mentioned earlier, plenty of resources and research are available exploring privacy and social media. As researchers begin to investigate the various aspects of this huge domain, they are making progress, but large gaps in knowledge remain. Specifically missing is how the privacy of personal information is interrelated to the discussion using social media to spread dis/misinformation. This effort will address a piece of that puzzle.

Methods

To investigate "How does the use of private information by social media companies lead to the spread of disinformation or misinformation?" this paper undertakes a qualitative study utilizing Grounded Theory to examine several instances of the spread of MIDI. Because research into social media is relatively new and accepted theories are rare, Grounded Theory offers an excellent tool to help discover commonalities and develop theory. As Glaser and Strauss point out, Grounded Theory offers outstanding theory generation, especially in new and emerging fields (Glaser and Strauss 2010).

The data used by this study consists of three well-developed examples of the spread of MIDI caused by social media corporations using private information. Descriptions of each instance will begin with some general background information to inform the reader, followed by discussing the financial incentive of the company. This is important: as stated earlier, the algorithms used by social media corporations are designed to keep users on their electronic devices to generate ad revenue. More importantly, the longer an individual is on their device using a social

media application, the more personal data is generated, captured, and ultimately used by the social media corporation. Thus, intimately tying financial incentives to collecting personal information. After discussing the financial incentives, the section will conclude with a discourse of the outcomes.

The researchers collected information regarding the individual instances from a variety of sources, including mainstream news sources and recent journal articles. They then compared the different articles to ensure consistency of the information. Each instance was then reviewed numerous times, using a cyclical methodology, to tease out common themes and data points. The ultimate goal was to determine if there were some commonalities between the different instances that could demonstrate social media companies using personal information to spread MIDI. The cases consist of the Cambridge Analytica case, the Russian interference in the 2016 presidential election, and the insurrection attempt on January 6, 2020.

Cambridge Analytica

Background

The political manipulation enabled by loose data privacy in the Cambridge Analytica incident pushed users into different worlds in an extraordinary way. Cambridge Analytica worked for the Trump presidential campaign, using unique advertising theories combined with incredibly personal datasets to shape their part of his campaign strategy (Rosenberg, Confessore, and

Cadwalladr 2018). The company used the concept of "micro-targeting" to profile individual users and target ads to specific individuals based on these psychological profiles across the whole of the United States (ur Rehman 2019). The personality profile application they used had access to users' Facebook data and data from all their connections even though none of the connections agreed to share that data (ur Rehman 2019). The process used by Cambridge Analytica involved users completing the "thisisyourdigitallife" quiz (2019). Cambridge Analytica then recorded data from the quiz, the user's profile, and scraped the profiles of all the user's connections (2019). At no point did this request permission or acceptance from the user's connections for their data, taking data from up to 87 million users (2019). Cambridge Analytica targeted users with ads specific to their profile (2019). According to a previous Cambridge Analytica employee, the company based ad targeting around 253 predictions about the user's personality (Hern 2018). Cambridge Analytica crafted these ads and messages using "dog whistles" to create ads that meant something different to each target group, swapping messages around to have one effect on person X and the opposite effect on person Y and suppress voting intentions (2018). The former employee, Christopher Wylie, also stated that Cambridge Analytica "absolutely" made use of fake news in their tactics (BBC News 2018). Because of the targeted ads created with unethically accessed personal information, one can interpret that the campaign essentially showed the voters two different sets of facts about candidates to manipulate their votes. The MIDI came into play with the content of the ads as they preyed on specific users' fears (as understood from the 253 data points) to show them false information with the most impact on that user (Heawood 2018; BBC News 2018; ur Rehman 2019). Cambridge Analytica presented something extra insidious by being less obvious than fake news lacking personalization (Heawood 2018). Nonetheless, Cambridge Analytica's operation did not last forever, and the company filed for bankruptcy in spring 2018 (Confessore and Rosenberg 2018). A U.S. Senate hearing on the matter ended with an unprecedented $5 billion fine on Facebook from the Federal Trade Commission (Hu 2020).

Financial Incentive

Although, on the surface, social media offers free connection with friends, family, and acquaintances, therein lies a strong profit motivation to connect users' data with advertisers. In 2018, when Cambridge Analytica came to light, Facebook had a worth of over $540 billion (ur Rehman 2019). Out of the $85.965 billion in revenue in 2020, $ 84.169 billion (over 97%) came from advertising ("Facebook - Financials" 2021). Based on those numbers, one can assume a significant incentive for Facebook to use data for advertising, regardless of privacy violations, since greed makes a powerful motivator (Crusius, Thierhoff, and Lange 2021). Those privacy violations ripen opportunities for exploitation via political manipulation, as seen with Cambridge Analytica's influence on the Trump campaign.

Outcome/Responses

Responses to Cambridge Analytica varied considerably, but several surveys showed it did affect public opinions on data privacy. A small Israeli survey of 51 adults (half before and half after Cambridge Analytica) noted a shift from viewing data privacy as a commodity to trade for services in the online world or a fundamental right to something one must give up to participate in the online world and something that regulation cannot enforce (Afriat et al. 2021). Another US-based in-depth interview study of 10 undergraduate students found that although none chose to leave Facebook permanently, several reported a reduction in Facebook use (Brown 2020). A New York Times article from a few weeks before the Senate hearing interviewed several people who chose to leave Facebook permanently – a decision many users found difficult but necessary to protect their privacy (Hsu 2018). Although these three sources might have predicted an adverse change in users for Facebook, the financial disclosures show a continuously increasing profit since 2018 ("Facebook - Financials" 2021). Nevertheless, Cambridge Analytica is not the only way Facebook's use of data has encouraged false information.

Russian Exploitation of Engagement Algorithms for Election Interference

Background

During the 2016 election campaign cycle, Russian operatives used Facebook to target American voters with over 129 phony political events (Timberg and Dwoskin 2018). Probably the most interesting one Facebook exposed details about was a double rally, planned and advertised by two separately controlled Russian Facebook groups (2018). The Heart of Texas "Stop Islamization of Texas" rally at the same time and place as their planned "Save Islamic Knowledge" rally hosted by the Russian-controlled group United Muslims of America occurred on May 21, 2016 (2018). Over 12,000 people viewed one paid promotion, and 2,700 viewed the other (Bertrand 2017). Details about the event came from the Senate Intelligence Committee (2017). The Heart of Texas page garnered over 225,000 followers, allowed the group to purchase ads for their events, and had real people show up to the events before Facebook shut it down (2017).

During the 2016 election cycle, Facebook hosted more than 3,000 ads about the U.S. presidential election created and purchased by the Russian Internet Research Agency (Isaac and Shane 2017). The Internet Research Agency, or IRA, spreads Kremlin-linked disinformation and propaganda for Russia, with various political aims linked to Russian goals (Isaac and Shane 2017; Permanent Select Committee on Intelligence 2018). Over 10 million people saw these ads, with 44% seen before election day and the remainder before congress requested to see the ads in 2017 (Isaac and Shane 2017). This highly coordinated, nation-state campaign demonstrates how little vetting Facebook gives to those who purchase ads from the company.

Financial incentive

The revenue generation mechanisms of social media are prone to spreading disinformation by relying on user engagement and viral content to garner revenue through advertising (O'Neil 2021; Walker, Mercea, and Bastos 2019). User engagement metrics come from the amount of time users are active on Facebook, the content they look at, and what they share, like, or comment on (Kim 2021). Facebook provides a particularly potent example of how these metrics feed into disinformation and misinformation spreading through their use of engagement algorithms. The metric of user engagement forms a particular threat because it offers financial incentives through advertising to psychologically exploit users and keep them looking at the platform — since engagement increases when users have an emotional reaction to content, the algorithm feeds users content they react to emotionally (Unger 2020). The algorithm personally curates each user's feed based on the user's viewing habits, interactions, and engagements with the platform (Oremus et al. 2021). Facebook's data collection does not stop with platform interactions either. The company uses cross site tracking to see what users purchase,

login to, view, and interact with on their browser (O'Flaherty 2021). Depending on user settings, or if they have the app on their phone, this tracking may even extend beyond the browser to cross-application tracking (2021). User settings, browser type, and whether or not users download the Facebook application can impact this part of the data collection. However, per Facebook's data use policy, the only people who can opt out of any private data processing are users protected by the European Union's General Data Protection Regulation, but these protections are unavailable to anyone else (Facebook 2022). This means all other users are subject to their data functioning as part of the engagement algorithm. Taking this data to target negative emotions appears to be the most effective method of increasing engagement. As noted by a recent Washington Post article, "The theory was simple: Posts that prompted lots of reaction emoji tended to keep users more engaged, and keeping users engaged was the key to Facebook's business" (Oremus and Merrill 2021, 2). 98% of Facebook's revenue generation in 2020 came through advertising (SEC 2021). Keeping users on the platform means they view more ads and increase revenue.

Outcome/Responses

"The company's data scientists confirmed in 2019 that posts that sparked angry reaction emoji were disproportionately likely to include misinformation, toxicity and low-quality news." (Oremus and Merrill 2021, 2). "Inducing reliance on emotion resulted in greater belief in fake (but not real) news stories compared to a control or to inducing reliance on reason" (Martel, Pennycook, and Rand 2020, 1). As seen by these sources, anger and misinformation based on user data form the backbone of Facebook's engagement algorithm, which is how they stay in business. Besides the examples of fake news spreading, the most significant outcome is users' emotional reaction to fake news.

January 6th

Background

The attempted insurrection on January 6th, 2021, and the associated social media activity renewed interest and attention to how social media affects politics. A New York Times article from a few weeks after the attack noted that approximately 140 police officers were injured, Officer Brian Sicknick died, and a total of four other cops committed suicide after the events (Schmidt and Broadwater 2021; Cameron 2022). Consider how one news article noted that former President Trump's Tweets fanned the flames of this attempt, and another noted evidence of plans to attack placed visibly on social media (Dilanian and Collins 2021; Miller, Jaffe, and Nakhlawi 2021). This case study starts with Twitter's financial incentives for pushing MIDI content to users, how the algorithm works with private user data, and then explores how this impacted the insurrection.

Financial Incentive

In 2016, Twitter changed the way its feed works from a reverse chronological timeline to one created by a ranking, or engagement, algorithm (Koumchatzky and Andryeyev 2017). This new algorithm functioned similarly to how Facebook's algorithm works by gathering user-specific data to create a personalized platform for each individual (2017). Some of this user data is the more public data such as likes, comments, and posts (2017). Other pieces of this data come from "[t]weets you found engaging in the past, how often and how heavily you use Twitter" and data that most users may not even realize Twitter is collecting – much less how they use it to manipulate what things the user sees first (2017). Unfortunately, this deep learning algorithm causes a similar problem to that seen in the Facebook engagement algorithm case study. A 2018 article from the Brookings Institute put it succinctly, saying, "we're more likely to react to content that taps into our existing grievances and beliefs, inflammatory tweets will generate quick engagement" (Meserole 2018, para. 10). A Cornell study estimated Twitter had over 10.64 million tweets and 35.84 retweets regarding election fraud claims (Abilov et al. 2021). A Pew Research study found that the average top 25% of Tweeters receive one retweet per month (Mcclain et al. 2021). Based on the Cornell numbers, each tweet about election fraud received roughly 3.36 retweets (Abilov et al. 2021). Retweets accounted for roughly 77% of engagement regarding election fraud

(2021). On average, only 49% of tweets shared on Twitter are retweets, with an additional 33% of engagement coming from replies (Mcclain et al. 2021). Since the Cornell study numbers do not include replies, this would suggest a higher-than-average secondary engagement for election MIDI content (Abilov et al. 2021). As mentioned earlier, increased engagement drives advertising which forms 89% of Twitter's revenue, according to their fiscal year 2021 financial disclosures (Twitter 2022a). If the election fraud tweets did indeed have a higher engagement rate, then Twitter would have a financial incentive to promote them above other tweets. Twitter's algorithm, based on engagement, would also be more likely to promote such tweets.

Twitter offers extremely limited options for opting out of personalized ads using such user data. This year, a blog post from the company stated, "Opting out of Twitter's interest-based ads won't stop you from seeing Twitter ads altogether. For example, you may still see ads on Twitter that are personalized based on other information, including what you Tweet, who you follow, what type of phone you use, where you are, and the links you click on Twitter." (Twitter 2022b). A 2018 article from MIT found that advertising may be responsible for as much as 75% of fake news spread (Relihan 2018).

To summarize, private data leads to social media companies (Twitter in this case) figuring out what content creates maximum engagement and emotional responses, advertising reve-nue incentivizes the company to create maximum engagement, and maximum engagement means pushing MIDI content to users. Election-related MIDI is no exception to this cycle. In two examples, a member of the group the Proud Boys and a Florida leader of a similar group, the Oath Keepers, both stated they went to the Capitol on January 6th because Trump had told them to be there (Feuer 2022). One of Trump's tweets on January 5th praised those who had already traveled to the capital, stating, "Washington is being inundated with people who don't want to see an election victory stolen by emboldened Radical Left Democrats. Our Country has had enough, they won't take it anymore! We hear you (and love you) from the Oval Office. MAKE AMERICA GREAT AGAIN!" (Trump 2021).

Outcome/Responses

Shortly after the election, CISA (Cybersecurity and Infrastructure Security Agency) put out a joint statement declaring, "There is no evidence that any voting system deleted or lost votes, changed votes, or was in any way compromised" (CISA 2020). According to the Department of Homeland Security, "The November 3rd election was the most secure in American history" (Al-Arshani 2020). Yet a Cornell study collected over 7.6M tweets and 25.6M retweets from 2.6M users related to voter fraud claims, and by their estimates, that was only 60% of the data (Abilov et al. 2021). Former President Trump himself tweeted over 300 times about election fraud before Twitter suspended him a few days after the insurrection (Qiu 2020; Abilov et al. 2021). Twitter cited incitement of violence and fear of further incitement of violence among their reasoning (Twitter 2021). His Facebook posts on mail-in voting fraud received roughly 3.8 million interactions (Unger 2020). Twitter suspended only 7.8% of accounts promoting such MIDI at the time of the Cornell dataset collection (Abilov et al. 2021). YouTube still had the top ten election MIDI spreading channels, and all of the top videos were online and visible as of Jan 11 (2021). The study found that such claims, and the spread through social media, undermined confidence in the election and played a significant role in the occurrence of the January 6th insurrection (2021). This creates the same issue as seen with Facebook, where emotional reactions prompt engagement, engagement feeds ad revenue, and engagement is highest when the algorithm presents dis- or misinformation - creating a financial incentive to continue the behavior (Oremus and Merrill 2021; Unger 2020). Although it is impossible to know exactly how much of the engagement regarding the election fraud claims twitter's algorithm drove, it seems clear that it had an important impact in fomenting the required emotions and belief for people to show up to the capitol on January 6th with intent to stop election proceedings. This instance exemplifies how Facebook and Twitter's engagement algorithms spread false information about the election being stolen and the impact it had on January 6th.

Bias and Limitations

Because of the secretive nature of social media conglomerates and their proprietary algorithms, getting exact data and examples is extremely difficult. This study relies heavily on secondary sources to evaluate instances of companies' use of private information, which can cause problems and inconsistencies. Additionally, these instances used in this qualitative study have yet to be fully detailed and documented as new information is coming to light almost daily.

Results

Each case study demonstrated a different violation of data privacy. Cambridge Analytica relied on microtargeting from personal datasets and psychological profiles, including data taken from individuals' connections (friends) without permission (Rosenberg, Confessore, and Cadwalladr 2018; ur Rehman 2019). Facebook's engagement algorithm relies on increased engagement as tracked by user viewing habits, interactions, and engagements with the platform – essentially tracking every mouse click and movement (Oremus et al. 2021). This includes cross-site tracking, where Facebook tracks purchases, logins, time spent viewing articles clicked on from Facebook posts, and other browser interactions (O'Flaherty 2021). It may also include cross-app tracking, depending on user settings and whether or not the user has the Facebook app (2021). Twitter uses similar engagement data to Facebook, tracking clicks, views, location, type of phone, and whom the user follows without options to change that tracking within the platform (Twitter 2022b).

Conclusion

Combining the everyday use of social media as a political messaging platform with currently available (and accurate) profiling allows a new exploit for disinformation campaigns that harm democratic institutions (Krafft and Donovan 2020). Analyzing the evidence presented in these case studies indicates that if an informed electorate is a vital part of democracy and advertising revenue disincentivizes private companies to encourage correct information, democracy cannot continue to exist as it has in the past. For years politicians have used social media platforms to curate semi-personal messages aimed at large masses (Sobieraj et al. 2020). Although disinformation campaigns are nothing new, the methods enabled by data privacy violations (personal profiling) and the megaphones social media gives users present a unique threat to democracy through the political incivility caused by these uses of social media (Unger 2020).

At the time of this writing, the congressional January 6th committee is still uncovering new information, the COVID-19 pandemic has left lingering effects and a society uncertain of where it stands on mandatory vaccinations, polio may be making a comeback, and a new health threat, monkeypox, is emerging (Plummer 2022; Cheng 2022; January 6th Committee 2022; Haseltine 2022). Future research should consider these latest variables and updated information. Ideally, more information will also come to light about how Twitter, Facebook, YouTube, and other social media giants use their algorithms.

As seen in the case studies, after the platforms collate this data, they use it to personalize advertising, increase engagement, and increase advertising revenue. Frequently, this comes at the cost of spreading MIDI due to the emotional reaction false information causes, which leads to increased engagement. The fake news then increases political incivility, in some cases leading to civil unrest based on falsehoods, which negatively effects democracy.

Policy Implications

As mentioned earlier, the only exceptions Facebook applies to their data handling are for users who fall under the protection of the GDPR (Facebook 2022). The GDPR, or general data protection regulation, is a European Union (EU) regulation that advertises itself as the toughest in the world and includes hundreds of pages of requirements applicable to any organization targeting EU users or data related to EU citizens ("Data Protection under GDPR" 2021). The legislation offers consumer protections, including regulations regarding obtaining consent for data processing, what types of data companies can process, and data transfer ("Data Protection under GDPR" 2021). This forward-thinking legislation gives users a say in how their data is handled and prevents scenarios like the case studies discussed in this paper from occurring by stopping the problems at the source – the data.

As opposed to the protections in the EU, data privacy legislation in the United States is currently a patchwork effort of minimal federal efforts and some state laws, unlike the consumer protections offered in the European Union. Although data broker sounds like something from a sci-fi dystopia, it is currently a billion-dollar industry in the U.S. lacking necessary federal regulation regarding the collection or use of said data (Martin 2020). In the United States, California passed the California Consumer Privacy Act of 2018 governing knowledge about what data businesses are collecting, options to opt-out of data sharing, and the right to delete personal information the businesses collected ("California Consumer Privacy Act (CCPA)" 2018). Vermont also passed H.764 to give consumers the right to opt out of data collection and the right to know more about data brokers' collection practices ("Bill Status H.764 (Act 171)" 2021). Current federal regulations include the Health Insurance Portability and Account-

ability Act or HIPAA (which applies to healthcare data)("Health Information Privacy" 2015), and the Gramm-Leach-Bliley Act (applied to financial data) ("Gramm-Leach-Bliley Act" 2021). In comparison with other data privacy-conscious governments, the next logical step would involve a comprehensive legislative effort for consumer privacy protections at the federal level, like the GDPR.

In addition to regulating data privacy, congress should also consider adopting ethical AI (Artificial Intelligence) legislation to correct malicious, ill-informed or negatively consequential AI. Last year, the Department of Defense mandated five principles to ensure the ethical, legal, and safe use of AI in warfighting and business applications (Hicks 2021). These include ensuring AI is responsible, equitable, traceable, reliable, and governable (2021). In context of the social media algorithms and the harm they cause the most important of these is traceability which would allow both the company, and if congress mandated it, governmental regulatory bodies to understand how these algorithms work and ensure they do not cause harm for individuals or democracy as a whole.

References

Abilov, Anton, Yiqing Hua, Hana Matatov, Ofra Amir, and Mor Naaman. 2021. "VoterFraud2020: A Multi-Modal Dataset of Election Fraud Claims on Twitter." arXiv. http://arxiv.org/abs/2101.08210.

Abril, Danielle. 2019. "Mark Zuckerberg Calls Facebook a Free-Speech Zone as Critics Demand More Restrictions." Fortune. October 17, 2019. https://fortune.com/2019/10/17/facebook-ceo-mark-zuckerberg-freedom-of-expression-speech/.

Afriat, Hagar, Shira Dvir-Gvirsman, Keren Tsuriel, and Lidor Ivan. 2021. "'This Is Capitalism. It Is Not Illegal': Users' Attitudes toward Institutional Privacy Following the Cambridge Analytica Scandal." *The Information Society* 37 (2): 115–27. https://doi.org/10.1080/01972243.2020.1870596.

Al-Arshani, Sarah. 2020. "The Department of Homeland Security Breaks from Trump and His Baseless Claims of Election Fraud, Calling This Year's Presidential Race 'the Most Secure in American History.'" Business Insider. November 12, 2020. https://www.businessinsider.com/dhs-breaks-from-trump-2020-election-most-secure-in-history-2020-11.

Asongu, Simplice A, and Stella-Maris I Orim. 2019. "Terrorism and Social Media: Global Evidence." *AGDI Working Paper*, 30.

Avaaz. 2021. "A Shot in the Dark: Researchers Peer under the Lid of Facebook's 'Black Box,' Uncovering How Its Algorithm Accelerates Anti-Vaccine Content." https://avaazimages.avaaz.org/fb_algorithm_antivaxx.pdf.

Baptist, Jeffrey, and Julian Gluck. 2021. "The Gray Legion: Information Warfare Within Our Gates." *Journal of Strategic Security* 14 (4): 37–55. https://doi.org/10.5038/1944-0472.14.4.1928.

Barnes, Julian E. 2021. "Russian Disinformation Targets Vaccines and the Biden Administration." *The New York Times*, August 5, 2021. https://www.nytimes.com/2021/08/05/us/politics/covid-vaccines-russian-disinformation.html.

BBC News. 2018. "'Cambridge Analytica Planted Fake News.'" *BBC News*, March 20, 2018. https://www.bbc.com/news/av/world-43472347.

Beigi, Ghazaleh. 2020. "Protecting User Privacy with Social Media Data and Mining." Dissertation, Arizona, USA: Arizona State University.

Bertrand, Natasha. 2017. "Russia Organized 2 Sides of a Texas Protest and Encouraged 'Both Sides to Battle in the Streets.'" Business Insider. November 1, 2017. https://www.businessinsider.com/russia-trolls-senate-intelligence-committee-hearing-2017-11.

"Bill Status H.764 (Act 171)." 2021. 2021. https://legislature.vermont.gov/bill/status/2018/H.764.

Brannon, Valerie C. 2018. "Free Speech and the Regulation of Social Media Content." R45650. Washington D.C.: Congressional Research Service. https://doi.org/10.1201/b22397-4.

Brown, Allison J. 2020. "'Should I Stay or Should I Leave?': Exploring (Dis)Continued Facebook Use After the Cambridge Analytica Scandal." *Social Media + Society* 6 (1): 2056305120913884. https://doi.org/10.1177/2056305120913884.

"California Consumer Privacy Act (CCPA)." 2018. State of California - Department of Justice - Office of the Attorney General. October 15, 2018. https://oag.ca.gov/privacy/ccpa.

Cameron, Chris. 2022. "These Are the People Who Died in Connection With the Capitol Riot." *The New York Times*, January 5, 2022, sec. U.S. https://www.nytimes.com/2022/01/05/us/politics/jan-6-capitol-deaths.html.

Cascini, Fidelia, Ana Pantovic, Yazan A. Al-Ajlouni, Giovanna Failla, Valeria Puleo, Andriy Melnyk, Alberto Lontano, and Walter Ricciardi. 2022. "Social Media and Attitudes towards a COVID-19 Vaccination: A Systematic Review of the Literature." *EClinicalMedicine* 48 (June): 101454. https://doi.org/10.1016/j.eclinm.2022.101454.

CCDH. 2021. "The Disinformation Dozen: Why Platforms Must Act on Twelve Leading Anti-Vaxxers." https://252f2edd-1c8b-49f5-9bb2-cb57bb47e4ba.filesusr.com/ugd/f4d9b9_b7cedc0553604720b7137f8663366ee5.pdf.

Cheng, F.K. 2022. "Debate on Mandatory COVID-19 Vaccination." *Ethics, Medicine, and Public Health* 21 (April): 100761. https://doi.org/10.1016/j.jemep.2022.100761.

CISA. 2020. "Joint Statement from Elections Infrastructure Government Coordinating Council & the Election Infrastructure Sector Coordinating Executive Committees." November 12, 2020. https://www.cisa.gov/news/2020/11/12/joint-statement-elections-infrastructure-government-coordinating-council-election.

Confessore, Nicholas, and Matthew Rosenberg. 2018. "Cambridge Analytica to

File for Bankruptcy After Misuse of Facebook Data." *The New York Times*, May 2, 2018, sec. U.S. https://www.nytimes.com/2018/05/02/us/politics/cambridge-analy tica-shut-down.html.

Cotter, Kelley, Julia R. DeCook, and Shaheen Kanthawala. 2022. "Fact-Checking the Crisis: COVID-19, Infodemics, and the Platformization of Truth." *Social Media + Society* 8 (1): 205630512110690. https://doi.org/10.1177/20563051211069048.

Crusius, Jan, Josephine Thierhoff, and Jens Lange. 2021. "Dispositional Greed Pre-dicts Benign and Malicious Envy." *Personality and Individual Differences* 168 (Jan-uary): 110361. https://doi.org/10.1016/j.paid.2020.110361.

"Data Protection under GDPR." 2021. Your Europe. 2021. https://europa.eu/you reurope/business/dealing-with-customers/data-protection/data-protection-gdpr/index_en.htm.

De Vynck, Gerrit, and Rachel Lerman. 2021. "Facebook and YouTube Spent a Year Fighting Covid Misinformation. It's Still Spreading." *Washington Post*, July 22, 2021. https://www.washingtonpost.com/technology/2021/07/22/facebook-youtu be-vaccine-misinformation/.

Dilanian, Ken, and Ben Collins. 2021. "Feds Aren't Using Posts about Plans to Attack the Capitol as Evidence." NBC News. April 20, 2021. https://www.nbcnews.com/politics/justice-department/we-found-hundreds-posts-about-plans-attack-capitol-why-aren-n1264291.

"Disinformation Nation: Social Media's Role in Promoting Estremism and Misin-formation." 2021. Washington D.C.

Dubois, Elizabeth, and Grant Blank. 2018. "The Echo Chamber Is Overstated: The Moderating Effect of Political Interest and Diverse Media." *Information, Communi-cation & Society* 21 (5): 729–45. https://doi.org/10.1080/1369118X.2018.1428656.

Facebook. 2022. "Facebook Data Policy." January 4, 2022. https://m.facebook.com/privacy/explanation/?_se_imp=1B8bi41uKqU3YqgfM.

"Facebook - Financials." 2021. 2021. https://investor.fb.com/financials/default.aspx.

Fallis, Don. 2015. "What Is Disinformation?" *Library Trends* 63 (January): 401–26. https://doi.org/10.1353/lib.2015.0014.

Feuer, Alan. 2022. "Three Characters at the Heart of an Unsettling Jan. 6 Narrative." *The New York Times*, June 10, 2022, sec. U.S. https://www.nytimes.com/2022/06/09/

us/politics/jan-6-proud-boys-capitol-police.html.

Franz, Daschel, Heather Elizabeth Marsh, Jason I. Chen, and Alan R. Teo. 2019. "Using Facebook for Qualitative Research: A Brief Primer." *Journal of Medical Internet Research* 21 (8): e13544. https://doi.org/10.2196/13544.

George, Daniel R., Liza S. Rovniak, and Jennifer L. Kraschnewski. 2013. "Dangers and Opportunities for Social Media in Medicine." *Clinical Obstetrics and Gynecology* 56 (3): 10.1097/GRF.0b013e318297dc38. https://doi.org/10.1097/GRF.0b013e318297dc38.

Glaser, Barney G., and Anselm L. Strauss. 2010. *The Discovery of Grounded Theory: Strategies for Qualitative Research*. 5. paperback print. New Brunswick: Aldine Transaction.

Goldstein, Neal D. 2021. "Misinformation." *American Journal of Public Health* 11 (2). https://www.proquest.com/docview/2486203133.

"Gramm-Leach-Bliley Act." 2021. Federal Trade Commission. 2021. https://www.ftc.gov/tips-advice/business-center/privacy-and-security/gramm-leach-bliley-act.

Haseltine, William A. 2022. "There May Be A New Polio Epidemic On Its Way- If So, What We Can Do: Part III." Forbes. June 28, 2022. https://www.forbes.com/sites/williamhaseltine/2022/06/28/there-may-be-a-new-polio-epidemic-on-its-wayif-so-what-we-can-do-part-iii/.

"Health Information Privacy." 2015. Text. HHS.Gov. August 26, 2015. https://www.hhs.gov/hipaa/index.html.

Heawood, Jonathan. 2018. "Pseudo-Public Political Speech: Democratic Implications of the Cambridge Analytica Scandal." *Information Polity* 23 (4): 429–34. https://doi.org/10.3233/IP-180009.

Hern, Alex. 2018. "Cambridge Analytica: How Did It Turn Clicks into Votes?" *The Guardian*, May 6, 2018, sec. News. https://www.theguardian.com/news/2018/may/06/cambridge-analytica-how-turn-clicks-into-votes-christopher-wylie.

Hicks, Kathleen. 2021. "Implementing Responsible Artificial Intelligence in the Department of Defense." Department of Defense.

Houston, J. Brian, Joshua Hawthorne, Mildred F. Perreault, Eun Hae Park, Marlo Goldstein Hode, Michael R. Halliwell, Sarah E. Turner McGowen, et al. 2015.

"Social Media and Disasters: A Functional Framework for Social Media Use in Disaster Planning, Response, and Research." *Disasters* 39 (1): 1–22. https://doi.org/10.1111/disa.12092.

Hsu, Tiffany. 2018. "For Many Facebook Users, a 'Last Straw' That Led Them to Quit - The New York Times." March 21, 2018. https://www.nytimes.com/2018/03/21/technology/users-abandon-facebook.html.

Hu, Margaret. 2020. "Cambridge Analytica's Black Box." *Big Data & Society* 7 (2): 2053951720938091. https://doi.org/10.1177/2053951720938091.

Hunter, Gary L., and Steven A. Taylor. 2019. "The Relationship between Preference for Privacy and Social Media Usage." *Journal of Consumer Marketing* 37 (1): 43–54. https://doi.org/10.1108/JCM-11-2018-2927.

Isaac, Mike, and Scott Shane. 2017. "Facebook's Russia-Linked Ads Came in Many Disguises." *The New York Times*, October 2, 2017, sec. Technology. https://www.nytimes.com/2017/10/02/technology/facebook-russia-ads-.html.

January 6th Committee. 2022. "Committee Activity." Select Committee to Investigate the January 6th Attack on the United States Capitol. July 21, 2022. https://january6th.house.gov/committee_activity.

Kim, Jiyoun. 2021. "The Meaning of Numbers: Effect of Social Media Engagement Metrics in Risk Communication." *Communication Studies* 72 (2): 195–213. https://doi.org/10.1080/10510974.2020.1819842.

Koumchatzky, Nicolas, and Anton Andryeyev. 2017. "Using Deep Learning at Scale in Twitter's Timelines." May 9, 2017. https://blog.twitter.com/engineering/en_us/topics/insights/2017/using-deep-learning-at-scale-in-twitters-timelines.

Krafft, P. M., and Joan Donovan. 2020. "Disinformation by Design: The Use of Evidence Collages and Platform Filtering in a Media Manipulation Campaign." *Political Communication* 37 (2): 194–214. https://doi.org/10.1080/10584609.2019.1686094.

Lee, Klara, and Mack Willett. 2021. "Rightly | Giant Tech Companies Following Your Personal Data." Rightly. February 18, 2021. https://right.ly/our-views-and-opinions/the-privacy-policies-of-social-media-companies-how-do-they-use-your-data/.

Li, Wei, and Hao Li. 2013. "Misinformation." *International Economic Review,* 54 (1): 253–77.

Martel, Cameron, Gordon Pennycook, and David G. Rand. 2020. "Reliance on Emotion Promotes Belief in Fake News." *Cognitive Research: Principles and Implications* 5 (1): 47. https://doi.org/10.1186/s41235-020-00252-3.

Martin, Brittany A. 2020. "The Unregulated Underground Market for Your Data: Providing Adequate Protections for Consumer Privacy in the Modern Era." *Iowa Law Review* 105 (2): 865–900.

Mcclain, Colleen, Regina Widjaya, Gonzalo Rivero, and Aaron Smith. 2021. "2. Comparing Highly Active and Less Active Tweeters." *Pew Research Center: Internet, Science & Tech* (blog). November 15, 2021. https://www.pewresearch.org/int ernet/2021/11/15/2-comparing-highly-active-and-less-active-tweeters/.

Meserole, Chris. 2018. "How Misinformation Spreads on Social Media—And What to Do about It." *Brookings* (blog). May 9, 2018. https://www.brookings.edu/ blog/order-from-chaos/2018/05/09/how-misinformation-spreads-on-social-med ia-and-what-to-do-about-it/.

Miller, Greg, Greg Jaffe, and Razzan Nakhlawi. 2021. "A Mob Insurrection Stoked by False Claims of Election Fraud and Promises of Violent Restoration." *Washington Post*, January 9, 2021. https://www.washingtonpost.com/national-security /trump-capitol-mob-attack-origins/2021/01/09/0cb2cf5e-51d4-11eb-83e3-322644d82356_story.html.

O'Flaherty, Kate. 2021. "All The Ways Facebook Tracks You And How To Stop It." Forbes. May 8, 2021. https://www.forbes.com/sites/kateoflahertyuk/2021/05/08/ all-the-ways-facebook-tracks-you-and-how-to-stop-it/.

O'Neil, Cathy. 2021. "Analysis | Facebook's Algorithms Are Too Big to Fix." *Washington Post*, October 8, 2021. https://www.washingtonpost.com/business/facebooks-algorithms-are-too-big-to-fix/2021/10/08/63986860-282f-11ec-8739-5cb6a ba30a30_story.html.

Oremus, Will, Chris Alcantara, Jeremy B. Merrill, and Artur Galocha. 2021. "How Facebook Shapes Your Feed." *Washington Post*, October 26, 2021. https:// www.washingtonpost.com/technology/interactive/2021/how-facebook-algo rithm-works/.

Oremus, Will, and Jeremy B. Merrill. 2021. "Facebook Prioritized 'Angry' Emoji Reaction Posts in News Feeds - The Washington Post," October 26, 2021. https:// www.washingtonpost.com/technology/2021/10/26/facebook-angry-emoji-algo rithm/.

Permanent Select Committee on Intelligence. 2018. "Exposing Russia's Effort to Sow Discord Online: The Internet Research Agency and Advertisements | Permanent Select Committee on Intelligence." June 2018. https://intelligence.house.gov/social-media-content/.

Plummer, Robert. 2022. "Monkeypox: WHO Declares Highest Alert over Outbreak." *BBC News*, July 23, 2022, sec. Health. https://www.bbc.com/news/health-62279436.

Qiu, Linda. 2020. "Trump Has Amplified Voting Falsehoods in over 300 Tweets since Election Night." *The New York Times*, November 16, 2020, sec. Technology. https://www.nytimes.com/2020/11/16/technology/trump-has-amplified-voting-falsehoods-in-over-300-tweets-since-election-night.html.

Rehman, Ikhlaq ur. 2019. "Facebook-Cambridge Analytica Data Harvesting: What You Need to Know," 12.

Relihan, Tom. 2018. "Social Media Advertising Can Boost Fake News — or Beat It." MIT Sloan. December 19, 2018. https://mitsloan.mit.edu/ideas-made-to-matter/social-media-advertising-can-boost-fake-news-or-beat-it.

Rosenberg, Matthew, Nicholas Confessore, and Carole Cadwalladr. 2018. "How Trump Consultants Exploited the Facebook Data of Millions." *The New York Times*, March 17, 2018, sec. U.S. https://www.nytimes.com/2018/03/17/us/politics/cambridge-analytica-trump-campaign.html.

Schmidt, Michael S., and Luke Broadwater. 2021. "Officers' Injuries, Including Concussions, Show Scope of Violence at Capitol Riot." *The New York Times*, February 12, 2021, sec. U.S. https://www.nytimes.com/2021/02/11/us/politics/capitol-riot-police-officer-injuries.html.

SEC. 2021. "Facebook FORM 10-K 2020." https://www.sec.gov/ix?doc=/Archives/edgar/data/1326801/000132680121000014/fb-20201231.htm.

Sethi, Umong. 2013. "Social Media - A Tool for the Military." *Centre for Land Warfare Studies*, 5.

Smith, Douglas, Trinity Leonis, and S. Anandavalli. 2021. "Belonging and Loneliness in Cyberspace: Impacts of Social Media on Adolescents' Well-Being." *Australian Journal of Psychology* 73 (1): 12–23. https://doi.org/10.1080/00049530.2021.1898914.

Sobieraj, Sarah, Gina M. Masullo, Philip N. Cohen, Tarleton Gillespie, and Sarah

J. Jackson. 2020. "Politicians, Social Media, and Digital Publics: Old Rights, New Terrain." *American Behavioral Scientist* 64 (11): 1646–69. https://doi.org/10.1177/0002764220945357.

Sufi, Fahim, and Musleh Alsulami. 2022. "A Novel Method of Generating Geospatial Intelligence from Social Media Posts of Political Leaders." *Information* 13 (3): 120. https://doi.org/10.3390/info13030120.

Thompson, Cadie. 2015. "What You Really Sign up for When You Use Social Media." CNBC. May 27, 2015. https://www.cnbc.com/2015/05/20/what-you-really-sign-up-for-when-you-use-social-media.html.

Timberg, Craig, and Elizabeth Dwoskin. 2018. "Russians Got Tens of Thousands of Americans to RSVP for Their Phony Political Events on Facebook." *Washington Post*, January 25, 2018. https://www.washingtonpost.com/news/the-switch/wp/2018/01/25/russians-got-tens-of-thousands-of-americans-to-rsvp-for-their-phony-political-events-on-facebook/.

Tourani, Nazanin. 2022. "Thriving in a Shifting Landscape: Role of Social Media in Support of Business Strategy." *Asia Pacific Management Review*, February, 1–6. https://doi.org/10.1016/j.apmrv.2021.11.001.

Trump, Donald. 2021. "Search on Trump Twitter Archive." January 5, 2021. //www.thetrumparchive.com.

Twitter. 2021. "Permanent Suspension of @realDonaldTrump." January 8, 2021. https://blog.twitter.com/en_us/topics/company/2020/suspension.

———. 2022a. "Fiscal Year 2021 Twitter Annual Report." 2022. https://s22.q4cdn.com/826641620/files/doc_financials/2021/ar/FiscalYR2021_Twitter_Annual_-Report.pdf.

———. 2022b. "Your Privacy Options for Personalized Ads | Twitter Help." April 28, 2022. https://help.twitter.com/en/safety-and-security/privacy-controls-for-tailored-ads.

Unger, Wayne. 2020. "How Disinformation Campaigns Exploit the Poor Data Privacy Regime to Erode Democracy." *SSRN Electronic Journal*. https://doi.org/10.2139/ssrn.3762609.

University of Washington. 2021. "Library Guides: News: Fake News, Misinformation & Disinformation." October 2021. https://guides.lib.uw.edu/c.php?g=345925&p=7772376.

Walker, Shawn, Dan Mercea, and Marco Bastos. 2019. "The Disinformation Landscape and the Lockdown of Social Platforms." *Information, Communication & Society* 22 (11): 1531–43. https://doi.org/10.1080/1369118X.2019.1648536.

Wilczek, Bartosz. 2020. "Misinformation and Herd Behavior in Media Markets: A Cross-National Investigation of How Tabloids' Attention to Misinformation Drives Broadsheets' Attention to Misinformation in Political and Business Journalism | PLOS ONE." *PLOS ONE*, https://doi.org/10.1371/journal.pone.0241389, , November. https://journals.plos.org/plosone/article?id=10.1371/journal.pone.0241389.

Winegar, A. G., and C. R. Sunstein. 2019. "How Much Is Data Privacy Worth? A Preliminary Investigation." *Journal of Consumer Policy* 42 (3): 425–40. https://doi.org/10.1007/s10603-019-09419-y.

The Cognitive-Strategic Meaning of the Special Military Operation in Ukraine

Dr. Eugene A. Vertlieb

Translated from the Russian by Dennis T. Faleris

1. "Soviet communism contains the seeds of its own self-destruction and will eventually collapse due to internal weakness," said politician George Kennan, considered the "architect of the Cold War."[1] Two detonation waves contributed to the collapse of the USSR. *First*, the Kremlin conspiracy of "doctor-saboteurs" liquidated the "father of the peoples."[2] As a result of the coup d'état, secured by a couple of hundred tanks near the Kremlin and bombers on takeoff, the "Khrushchev Thaw" began.[3] Former militants of the OUN-UPA,[4] Bandera,[5]

---

1   "The U.S. State Department asked George F. Kennan to analyze Soviet communism and put forth ideas on how to deal with Stalin and communist Russia. In response, Kennan, an expert of Russian history and culture who lived in Moscow as a senior diplomat, drafted an 8,000-word telegram in which he presented the rationale for what would become the US policy of containment." https://study.com/academy/lesson/george-f-kennan-containment-in-the-cold-war.html

2   "The Doctors' Plot," [Russian "Дело врачей"—"Doctors' Case" or "Дело вредителей"—"Case of the Saboteur Doctors"] refers to an alleged conspiracy of prominent, mostly Jewish, doctors from Moscow to murder leading Soviet government and party officials. Many doctors were arrested and tortured to produce admissions, but after the death of "the father of the peoples" (Stalin), it was declared that the case had been a fabrication. https://en.wikipedia.org/wiki/Doctors%27_plot The Soviet dictator, Stalin, ruled from 1929 to 1953.

3   "The Thaw" [Russian "Оттепель"] refers to the Soviet period from the mid-1950s to the early 1960s when repression and censorship in the Soviet Union were eased under USSR Communist Party Chairman Nikita Khrushchev. The term was coined after Ilya Ehrenburg's 1954 novel "The Thaw" which was published in 1954, a year after Stalin's death. https://www.newworldencyclopedia.org/entry/Khrushchev_Thaw

4   "The Organization of Ukrainian Nationalists (OUN) was a Ukrainian ultranationalist political organization established in 1929... It split into two parts in 1940...one of which supported Stepan Bandera. In 1941, it declared an independent Ukrainian state in occupied Lviv while the region was under the control of Nazi Germany...to free Ukrainians from Russian oppression. In 1942, the OUN established the Ukrainian Insurgent Army (UPA) which carried out large-scale ethnic cleansing against Polish people...to pre-empt Polish efforts to re-establish Poland's pre-war borders. https://en.wikipedia.org/wiki/Organization_of_Ukrainian_Nationalists

5   "Stepan Bandera was the leader of the revolutionary faction of the Organization of Ukrainian Nationalists, which, along with its partisan army—the Ukrainian Insurgent Army— strove to eliminate all ethnically non-Ukrainian elements from Ukrainian soil (including Jews, Russians, Poles, Gypsies, etc.) and, for a certain period of time, collaborated with the Germans in the hope of achieving this goal." See (2015, January 29) "The Success of Russia's Propaganda: Ukraine's "Banderovtsy.'" *Cambridge Globalist.* http://cambridgeglobalist.org/?p=573

doi: 10.18278/gsis.8.1.9

Vlasov[6] and other collaborators were amnestied. As stated in a recently declassified document, "the number of OUN members increased tenfold and tension increased many times throughout Ukraine." These are the origins of the Maidan.[7] The foundations of the Red Empire were shaken. The newest chapter in the history of Russia is one characterized by suicidal upheaval.

2. *Second,* the Belovezhskaya conspiracy,[8] perpetrated by the highest-level *nomenklatura,* decided the fate of the Soviet Union itself. It ceased to exist—"it committed suicide" although suicidal tendencies had not been observed in the state. On the contrary, the USSR was quite confidently keeping itself afloat. Yuri Andropov[9] strove to adapt the Chinese practice of Deng Xiaoping with its "free economic zones"— the experience of communism co-existing with technological reform.

And Mikhail Gorbachev quite confidently branded the "so-called democrats" who were "preparing a coup d'état" and hatching insidious plans for "the fragmentation of our great multinational state."

3. But actions speak louder than words. The state system made fatal mistakes. As soon as the GDR was surrendered in 1989, the entire socialist camp fell. The fall of the "Iron Curtain" was facilitated by the departure from the "Brezhnev" USSR Constitution's Article 6 "on the leading and guiding role of the CPSU in Soviet society." This democratic correction in the construct of the foundation of the state had detrimental consequences for those in power.

4. With the removal of this fundamental "stone" from the foundation of the sovereign monolith, the smooth functioning of the state machine

---

6   "Andrey Andreyevich Vlasov was a Soviet Red Army general and Nazi collaborator... After being captured, he defected to Nazi Germany and headed the Russian Liberation Army. Upon being captured by Soviet forces...he was tortured, tried for treason, and hanged." https://en.wikipedia.org/wiki/Andrey_Vlasov

7   "The Maidan Uprising was a wave of demonstrations and civil unrest in Ukraine, which began on 21 November 2013 with large protests in "Maidan Nezalezhnosti" [English "Independence Square"]... The protests were sparked by the Ukrainian government's sudden decision not to sign the European Union-Ukraine Association Agreement, instead choosing closer ties to Russia and the Eurasian Economic Union." https://en.wikipedia.org/wiki/Euromaidan

8   The Belovezha Accords, also known as the Minsk Agreement, was signed in 1991 by President Boris Yeltsin of Russia, Ukraine president Leonid Kravchuk, and Chairman of the Supreme Soviet of Belarus, Stanislav Shushkevich. They met secretly in a resort in Belovezhska Pushcha, just outside of Brest, Belarus. The Accords ended the Soviet Union and established the Commonwealth of Independent States. https://www.encyclopedia.com/history/encyclopedias-almanacs-transcripts-and-maps/belovezh-accords

9   Yuriy Andropov was selected as the new General Secretary of the Communist Part in the USSR following the death of long-time Soviet leader Leonid Brezhnev in 1982. https://www.history.com/this-day-in-history/yuri-andropov-assumes-power-in-the-soviet-union

was disrupted. The control flywheel was out of balance; the foundations of the state's system and its structural framework were doomed to fail. Soon Boris Yeltsin reported to the US Congress, "The communist idol that sowed social discord, enmity, and unprecedented cruelty everywhere on earth, and instilled fear in the human community, has collapsed. It has collapsed forever. *And I am here to assure you we will not allow it resurrect on our soil!*" However, the special operation "Our Crimea"[10] (such "polite people") and the current operation to demilitarize and denazify "Ukronazis" are a reversal of the trend towards Russian imperial behavior.

5. We should note that after the defeat in the Cold War, the Russian Federation pursued an *inertial* foreign policy. Only the force majeure of the prevailing circumstances prompted the top management to "do something." So the 2008 war between Georgia and Russia took place because of the Kremlin's indecision over annexing Abkhazia and South Ossetia. The killing of peacekeepers was just a casus belli. The neologism *"Krymnash"* (2014) is a response, using "soft power," to the ousting of the Russian fleet from the Black Sea. Then "everything fell asleep" until a new threat caused the Russian Federation to wake up in a cold sweat. It turned out that the "Ukry"[11] were getting ready to join NATO (and this would mean four minutes of missile flight time to reach the Kremlin!). And a battle broke out, a military special operation, in fact, to restore the status quo—NATO's disposition as of 1997 before the alliance moved Eastward. (By comparison, the cleansing of separatist "Ichkerism"[12] was called "restoring constitutional order.")

6. If the "Russian bear" had not been awakened, the Kremlin would hardly have gone the route of special operations. Politician Alexander Prokhanov[13] states that

---

10 "Our Crimea" [Russian "Крымнаш," transliterated "Krymnash"] is a neologism that came into being in 2014 to refer to operations leading to the annexation of Crimea by the Russian Federation. "Krymnash" is a Romanized word combination which translates to "Crimea-our." Operation Krymnash is, therefore, "Operation Our Crimea. Goble, Paul. (2015, June 10) "'Krymnash' Meme Part of Russian Society's Return to Late Soviet Times." *Euromaidanpress*. http://euromaidanpress.com/2015/06/10/krymnash-meme-part-of-russian-societys-return-to-late-soviet-times/

11 Although the term may refer to ancient tribes called the "Ukry" [Russian "Укры"] who were the forerunners of the modern-day Ukrainian people, another use of the term is to refer to Ukrainians in a derisive manner. It is used to reflect the attitude of the Russian ruling elite toward the people of Ukraine.

12 The Chechen Republic of Ichkeria is the unrecognized secessionist government of the Chechen Republic. The republic was proclaimed in 1991, which led to two wars with the Russian Federation. https://military-history.fandom.com/wiki/Chechen_Republic_of_Ichkeria

13 Aleksandr Prokhanov, Editor of Zavtra, criticized Putin as being too liberal in the early days, but became a loyal supporter after the 2014 annexation of Crimea... Zavtra, anticipating Russian Easter, pictured the special military operation in Ukraine as a Holy war and depicted putting car-

Vladimir Putin "was like a balance beam scale on which were teetering two containers each holding a different structure—a patriotic one and a liberal one. But in a single instant, the scale became unbalanced. The patriotic order of things got out of Putin's control as did the liberal one. *The President did not manage to make the long-awaited break-through in that period of time after Crimea—a development that would have melded the two structures. And each went its own way.* Within each, a confusing mess arose—a complex system of decay. Because of this, *Putin has no control over these two massive aspects of modern Russia...* which are developing each in its own way, and pretty chaotically."[14] In Ukraine, by contrast, the war revealed the preparedness of their society to wage organized resistance against the "Russkies"[15] and for consolidating the elites. One of the mistakes of the Russian army was

the expectation that it would "be greeted with flowers" in Ukraine, but this notion has long since been dispelled, according to ex-commander of the Airborne Troops Colonel General Vladimir Shamanov.[16] In Ukraine, many military schools have been preserved since Soviet times—schools which can be found in almost every region. This allowed the country to preserve the schooling which served as "the basis for reviving the army."

7. With the defeat of the Soviet Union at the anti-communist stage of the permanent Cold War between the West and Russia, society became fragmented; a "parade of sovereign states" filled the post-Soviet space. After all, Article 72 of the USSR Constitution allowed for the withdrawal of the republics from its structure "right up to allowing secession." Instead of a pragmatic Chinese transformation, Russia "defined itself" in a Leninist

---

rying the cross of the "Russian Cathedral" to the top of Golgotha. Editor Aleksandr Prokhanov splashed the headline "He is truly risen!" and a picture of Christ across the front page, accompanied by heavy artillery, a tank, and some smoking buildings. https://cepa.org/putin-the-risen-christ-russias-ultra nationalist-pin-up/ "Zavtra"is an ultranationalist broadsheet founded by Prokhanov. The current editor is his son, Andrey Fefelov. See Balmforth, Tom. (2014, August 17) "From The Fringes Toward Mainstream: Russian Nationalist Broadsheet Basks In Ukraine Conflict." *Radio Free Europe/Radio Liberty.* https://www.rferl.org/a/26534846.html

14 Here, Prokhanov implies that Putin failed to strike a balance between patriotism and liberalism following the "Our Crimea" operation. See Prokhanov, Aleksandr. (2019, November 20) "The Two Pine Trees of President Putin." *Zavtra.* In this blog article, Prokhanov uses "pine tree" as it is used in the folk saying "to get lost in a forest of three pine trees" ["заблудиться в трех соснах"], which is said of someone who is unable to find any solution to a simple problem or situation. https://zavtra. ru/blogs/dve_sosni_prezidenta_putina

15 The Russian text uses the term "Moskaly" [Russian "Москали"]—an ethnic slur that means "Russians" in Ukrainian, Polish, and Belarusian.

16 Col-Gen Vladimir Anatolyevich Shamanov headed the Russian Airborne Troops from 2009 to 2016. After his retirement, he became head of the State Duma Defense Committee.

way—a pure "golovotyap" move (to use the language of the satirist Saltykov-Shchedrin)[17]—when it decided to become independent *from itself* (insanely mistaking its Union republics for "colonies"). Having rejected the ideology of the "Soviet Union,"[18] the "foremen of perestroika,"[19] compensating for its absence, instead threw at the masses the [ideological] cry of the shopkeepers, "Get rich!"

8. Under this French motto "Enrichissez-vous!" the "Russian Federation, as a state corporation," returned to the feudalism of a pakhanate[20]-comprador stewardship of the country. This slogan of the French liberal politician Francois Guizot was used in 1925 by Bolshevik ideologue Nikolai Bukharin[21] so that reviving elements of capitalism during NEP[22] would help the poor, who were in critical need, to survive. Alas, the very poor in the Russian Federation are not decreasing in number. By the end of 2022, more than 20% of Russian citizens will fall below the poverty line.

---

17 *Golovotyap* is one who carelessly and stupidly conducts any business — a bungler. The word was coined In the 19th century by satirical writer Mikhail Yevgrafovich Saltykov-Shchedrin (nom du plume "Nikolay Shchedrin"). The word "golovotyap" was formed by adding the roots of the noun "head" (golovo) and the verb "to chop" (tyapat'").

18 The Russian text uses the popular slang for the former Soviet Union — "Sovka" [Russian "Совка"].

19 McCauley, Martin; Lieven, Dominic. "The Gorbachev era: perestroika and glasnost." *Britannica.* "When Gorbachev became head of the Communist Party in 1985, he launched perestroika ('restructuring')... to bring the Soviet Union up to par economically with the West... In 1987-88, he pushed through additional reforms... The consequences of this form of a semi-mixed economy... brought economic chaos to the country." https://www.britannica.com/place/Russia/The-Gorbachev-era-perestroika-and-glasnost

20 Young, Cathy. (2014, August/September) "Putin's New Old Russia." *Reason.* A "pakhanate" is a coinage derived from pakhan [Russian "пахан"], Russian underworld slang for a gang boss. https://iq.hse.ru/news/177743032.html

21 Nikolay Bukharin was one of the key Soviet politicians in the years immediately following the 1917 Bolshevik Revolution. He was a theoretician o the Communist movement during the revolutionary period and throughout the 1920s. In January 1937 he was accused of treason and died by firing squad in 1938. https://russiapedia.rt.com/prominent-russians/politics-and-society/nikolay-bukharin/

22 NEP or "New Economic Policy" [Russian: "НЭП" or "Новая Экономическая Политика"] refers to "the policy of the government of the Soviet Union, in effect from 1921 to 1928, that represented a temporary retreat from its previous policy of extreme centralization and doctrinaire socialism. [The policy] was viewed by the Soviet government as merely a temporary expedient to allow the economy to recover while the Communists solidified their hold on power." https://www.britannica.com/event/New-Economic-Policyx-Soviet-history Further, "NEP was characterized by a partial denationalization of property, privatization of many previously nationalized enterprises, an economic management system based on cost accounting, competition, and the introduction of the leasing of joint ventures... The NEP led to a rapid economic recovery. The economic interest that appeared among the peasants in the production of agricultural products made it possible to quickly saturate the market with food and overcome the consequences of the hungry years of 'war communism.'" https://www.krugosvet.ru/enc/istoriya/NOVAYA_EKONOMICHESKAYA_POLITIKA_NEP.html

9. A foreign example of socio-political transformation: "Perestroika" was based on aspects of the French Revolution. If you look for the roots of our perestroika, you can trace back to the French Revolution, and then to the Commune. "Gorby" himself suggested the presence of foreign patterns in his project to reform state systems. In "Perestroika" loomed the possibility of a *shootout*.[23] First, a "shootout" of the "colonies" of the USSR among themselves, and then among the genetic and civilizational brothers—the "Russkies" and the "Ukies." (I felt the tectonics as the drama developed in the geopolitical tragedy of the USSR already back in the 1980s as expressed in my article entitled "Genesis and the paths of political consciousness under Gorbachev: 'Perestroika'—Evolution or *Shooting before a Shootout,*"[24] published in the well-known Parisian magazine, "Continent."[25]) Suicidal feelings were to some extent felt by Putin when faced with the decisions made by the West: "He shot himself in the foot or in the head with their economic autos-da-fé."[26]

10. Director Stanislav Govorukhin called the liquidation of the USSR "The Great Criminal Revolution."[27]

---

23 There is a play on words in the original text by Dr. Vertlieb: The Russian word for "Shootout"— "перестрелка"—(transliterated into Roman text: "perestrelka,") sounds like "perestroika" [Russian "перестройка"].

24 Vertlieb, Evgeny Aleksandrovich. (1990, April) "Genesis and the Paths of Political Consciousness in the Era of Gorbachev. 'Perestroyka'—an evolution or shooting before the shootout." *Kontinent No. 63, pp. 189-223* [Original Russian: Вертлиб, Евгений. "Генезис и пути политического сознания при Горбачёве. 'Перестройка'--эволюция или пристрелка перед перестрелкой." Континент *No. 63, 189-223.*

25 "Kontinent" [Russian: "Континент"] was [founded as] an émigré dissident journal which focused on the politics of the Soviet Union and its satellites. The journal continues to be published in English and Russian by Russia House. https://en.wikipedia.org/wiki/Kontinent Elsewhere, the journal is characterized as a "Russian literary, journalistic and religious magazine. Positioned itself as a Christian-liberal publication... it was founded in 1974 in Paris as a publication of the 'third wave' of Russian emigration... Kontinent was published until 1992 under the editorship of [Russian author] Vladimir Maksimov. In 1992, the magazine began to appear in Moscow under the editorship of Igor Vinogradov." https://magazines-gorky-media.translate.goog/continent?_x_tr_sl=ru&_x_tr_tl=en&_x_tr_hl=en&_x_tr_pto=sc

26 *Auto-da-fé* refers to the ceremony for pronouncing judgment by the Inquisition which was followed by the execution of the sentence by secular authorities or, in broader terms, it refers to the burning of a heretic. https://www.merriam-webster.com/dictionary/auto-da-fé

27 See Govorukhin, S. (1993) The Great Criminal Revolution [Russian "Великая криминальная революция"]. *Andreyevsky Flag Publishers.* 126 Pages. ISBN 5856080262. Govorukhin produced a documentary film by the same name. For a brief description, see Douglas, Rachel. (1994, July 15) "Documentary film on Russian crime is presented in Washington." *Executive Intelligence Review*, Vol. 21 (Number 28), pp. 142-145: "Russian film director Stanislav Govorukhin produced a documentary film about the two years following the dissolution of the USSR in 1991...The movie -- also a book by Govorukhin -- exposes the ex-Communist officials who became Russia's *nouveaux riches* by getting a leg up on amassing wealth when Gaidar decontrolled prices, as well as the mafia

But there was a blessing in disguise: Although "Perestroika" was inspired by the "Paris Commune," it also brought with it an exit ramp leading to genuine democracy. But the Russian national self-movement was countered by *"comprador stability"* which plunged Russian society into cognitive dissonance—at least judging by the clash of the two opposing concepts of "patriotic authenticity" held by the head of Chechnya, Lieutenant-General Ramzan Kadyrov, and by Presidential Press Secretary Dmitry Peskov.[28]

11. Is it a "friendly" skirmish between them—two perfectionists of the Russian World—or is it already the confrontational phase in a pre-barricade split in society? "God forbid that we should see a Russian rebellion, senseless and merciless," Pushkin feared. In the current situation, it is not a "senseless" rebellion. And here's why. Judging by

the reaction to another incident (the recent attack on the Russian Federation ambassador in Warsaw),[29] General Kadyrov is firm and resolute: "Our attitude about this is rigid, and *we are asking the president to force all states that are not with us to get on their knees...* Why are they not taking action?" The question was stated point-blank and was intended to clarify the meaning of the Special Operation against Ukraine in terms of its expression of national values and the military policy being exercised. So "In the name of what//does the boot//trample the earth, creaking and rude?//Who is it above the sky of battles—//freedom?//God?//The ruble?" prophetically asks the poet and champion of the people, Vladimir Mayakovsky.[30]

12. *But are not government officials preparing the defeat of Russia?* ask

---

kingpins who became their fellow travelers to billionairehood through extortion rackets."

28 "...Kadyrov has allowed himself to openly criticize...the head of the Russian delegation in negotiations with Ukraine... Moreover, the Chechen leader has had public disputes with Kremlin spokesman Dmitry Peskov, whom Kadyrov accused of 'lack of patriotism.'" (2022, May 28) "Russian media report Putin's frustration with Chechen troops conduct in Ukraine. *Ukrayinska Pravda*. https://www.yahoo.com/video/russian-media-report-putins-frustration-133126616.html See also (2022, April 6) "Kadyrov's dispute with Peskov arouses political analysts' interest." *Caucasian Knot*. https://www.eng.kavkaz-uzel.eu/articles/58730/

29 Hassan, Jenifer (2022, May 9) "Russian ambassador doused with red paint by protesters in Poland." *The Washington Post*. https://www.washingtonpost.com/world/2022/05/09/russia-ambassador-poland-paint-victory-day-warsaw/

30 According to *Read and Write Verse* on line, the poem "In Answer!" [Russian "К ответу!"] "was written by Russian poet Vladimir Mayakovsky in 1917 and was intended to expose world capitalism. It is clear, however, that Mayakovsky was devoid of idealism, and perfectly understood what was happening around him... He still believed in the ridiculous promises that the land would go to the people and there would be equality in the country." https://pishi--stihi-ru.translate.goog/k-otvetu-mayakovskij.html?_x_tr_sl=ru&_x_tr_tl=en&_x_tr_hl=en&_x_tr_pto=sc

Ukrainian politician Oleg Tsarev[31] and Russian businessman Andrey Samokhin. After all, the current "elite" is categorically not ready for a mobilization "for the Motherland" and is waiting for the "controlled defeat" of Russia. The Special Operation, which has clearly entered the phase of protracted bloody military operations right up to possible direct clashes with NATO, requires completely different management, an economy, as well as the consciousness of the people and of the elite— different than it has been up to now and continues to be. "Mobilization" is the gathering of the will and spirit of all the component parts of the state into one fist. However, instead of this, there are still strange ambiguities on the part of those in power and their lackies. A huge army of officials either completely renounces the war ("and why do we need this war?") or maintains a deafening silence. They would like it such that "if there is a loss, then let it be in a controlled way, for example, in surrendering all or part of the liberated territories. Give back Crimea, but not immediately, for example,

postponing it for fifteen to twenty-five years. Defeat and sanctions, of course, will affect the standard of living: The standard of living in the country will go down, but this will not affect the elites much. And from this we see *caution that is turning into sabotage.* There is a significant lobby within the Russian leadership that allows "any geopolitical experiments" involving the country to be carried out. Although patriotism is fundamentally inherent in the Russian people, *the mandate of trust in the current regime is not eternal.* Is the Russian Federation not threatening to repeat at a new stage the well-known tragic formula of the last Russian Sovereign [Nicholas II]: "There is cowardice, deceit and betrayal all around?"[32] In the Russian establishment, there is a picture of exactly this *creeping betrayal.*

13. "In such a situation, the old elites are instinctively aligned toward destruction. They'll be better off if we lose. In fact, *it is better to lose right now*, before the confrontation has gone on too far. And so that it doesn't go too far, we should not welcome

---

31 Oleg Tsarev is a Ukrainian businessman, politician, and separatist leader who resided in eastern Ukraine. In July 2014, he became the speaker of the Parliament of "Novorossiya" [Russian "Новороссия" or "New Russia"], a confederation that included тне separatist Donetsk and Luhansk People's Republics, and served until its dissolution in 2015. He was sentenced in absentia to 12 years in prison in May 2022. https://en.wikipedia.org/wiki/Oleg_Tsaryov

32 "These words are taken from the diary (entry dated March 2, 1917) of Russian Emperor Nicholas II, made by him on the day of his abdication in favor of his brother, Mikhail Romanov... The emperor wrote, 'The bottom line is that in the name of saving Russia, keeping the army at the front and calm, this step must be taken... At one in the morning I left Pskov with a heavy feeling of what I had experienced. Around treason, and cowardice, and deceit. Ironically about betrayed, cowardly friends, about the wrong environment.'" http://www.bibliotekar.ru/encSlov/10/177.htm

or encourage a *recruitment of volunteers in Russia*, and we must ensure that *strikes are not made on critical Ukrainian infrastructure, and that proportional measures are not introduced in response to economic and financial sanctions, that is, that there are no retaliatory seizures of assets, and that debts to Western countries continue to be paid.* And we should refrain from *striking Ukrainian "decision-making centers."* Everything should be done to demonstrate to the enemy that we are ready to pay and repent, repent and pay but the implacable opposition resents this. Bewilderment among the people is multiplying, but clearly the people do not want to—or are unable to—coherently explain the power that is being exercised on their behalf even if it is being done through full-time government telepropagandists.

14. The *whole system* needs to be changed at this point. "In our country," writes economist Sergey Batchikov,[33] "as a result of 'liberal market reforms,' a unique—by world standards—economic model has been created and is functioning: One that is illegitimate from the point of view of public opinion; illegal from the point of view of sources governing the formation

of property rights; extremely inefficient and unproductive in terms of social reproduction; uncompetitive in the global and domestic markets; systemically managed and manipulated from the outside; socially unjust; extremely resource- and energy-intensive; primitive in its structure as well as suffering from a shrinking manufacturing capability and lacking in integration." The question arises: Is it possible to mobilize such a system to counter the military and civilizational challenge being thrown down before Russia by the West? Patriot Alexander Prokhanov summarizes: "*The state government assembled by Putin has rotted almost to the ground.* You need to start "reformatting" the system from the very top. And turning on the "operating mode" that the population must "get up, the country is huge"[34] would radically change the situation in society, and at the battlefront, and in the economy. But first of all, everyone should realize: *if not a victory, then destruction.*

15. The top authorities of the Russian Federation have the opportunity to defend the interests of the country using any means, but there is no will to do so. So they are unlikely to use nuclear weapons as a preventive

---

33   Sergey Anatolyevich Batchikov is a Russian economist, businessman, and social and political figure. He was, according to a 2016 online source, Director of the Center for Management Problems in Large Socio-Economic Systems of the International Research Institute for Management Problems. https://izborsk-club.ru/11017

34   This is the title and the first line of the Russian song "Get up, the country is huge!" The song was the creation of the poet Vasily Lebedev-Kumach and the composer Aleksandrov, and was composed on the night of 22 on to 23 June 1941 as a call to arms. Germany attacked the Soviet Union on 22 June 1941. https://en.topwar.ru/87896-vstavay-strana-ogromnaya.html

measure. Intelligence expert Yakov Kedmi[35] identifies the pain points found in the Russian decision-making and security system. He is convinced that Russia, possessing strategic trump cards, should play the game differently: "All that is needed is one or one-and-a-half volleys from a "multi-purpose" submarine with Zircons[36] and in ten minutes, about 50-60 power plants Britain would cease to exist. And *the whole of Great Britain would return to the Stone Age*." The Israeli analyst welcomes a change in the way that foreign policy decisions of the Russian Federation are made—from *compromise* at all costs to solving the issue. "Lavrov and Russian diplomacy have changed their tone." There is a departure from the "anesthesia" approach, which avoids causing any painful inconvenience to the West. The expert considers the excessive kindness and compassion of Russians to be a serious flaw "where it is absolutely not necessary to do this."

From Kedmi's point of view, *the military strategy of a counter- or defensive nuclear strike, which has been adopted in Russian military doctrine as well as in the foundations of the Russian Federation state policy regarding nuclear deterrence, should be revised.* A counter nuclear strike has one drawback: "*It does not stop nuclear missiles that fly in your direction. It prevents only one thing—a preventive strike.*" It deprives the enemy of the opportunity to destroy or cause irreparable damage to your state, while a counterstrike only scares your opponent. What if they're not scared? "What if they are half-idiots or what if they have miscalculated? And the missiles that they sent your way are in the air. The first task of any state is to protect itself. To protect itself by destroying the enemy."[37] Yakov Kedmi is sure that a real war is going on against Russia, not a hybrid one—a war aimed at the complete destruction of Russia as a state by 2025. And given this timetable,

35   Yakov Kedmi (also known as Yasha Iosifovich Kazakov) was born in Moscow, but denounced Soviet citizenship in 1968 and emigrated to Israel. In the 2010s, Kedmi became a regular participant in propaganda shows on Russian TV where he has urged Russia to start a war with the US, defended Stalin, and invariably defends the Kremlin's point of view on foreign policy. He regularly accuses the Ukrainian leadership of Nazism and Ukrainian society of anti-Semitism. See "Kedmi Yakov." *Putin's List, Free Russia Forum.* https://www.spisok-putina.org/en/personas/kedmi-2/

36   Dangwal, Ashish. (2022, May 28) "Russia's Zircon Hypersonic Missile Fired From Admiral Gorshkov Hits Target Bang-On 1000 Km Away In The White Sea." *Eurasian Times.* Zircon [Russian "Циркон"] is a Russian-produced anti-ship hypersonic cruise missile. In October 2021, Russia's nuclear-powered submarine Severodvinsk launched Zircon hypersonic missiles from surface and submerged positions in the White Sea for the first time. https://eurasiantimes.com/russia-developing-a-new-coastal-defense-system-for-its-zircon/

37   See, for example, (2022, May 6) "Kedmi Yakov Iosifovich Interview." *Evening with Vladimir Solovyov. Polit Analyst: Opinions and statements of famous political scientists.* https://polit--analitic-ru.translate.goog/category/06-maya-gost-peredachi-kedmi-ya-i/?_x_tr_sl=ru&_x_tr_tl=en&_x_tr_hl=en&_x_tr_pto=op,sc

the United States is operating using all possible means of destruction. Therefore, today, in order to win, Russia, according to Kedmi, must change all of its "last-year" concepts and strategies and *adapt them to a new period of military confrontation with the West—the 2022-2023 period*. Russia has no right not to win this fight.

16. The Russian victory is making progress in microsteps. The remnants of the USSR are destroying each other. According to the calculation of liberal observer Dmitry Demushkin,[38] Russia, in the three months of the Special Operation, has: lost its gold and foreign exchange reserves; become the target of every possible sanction; dropped out of world institutions; managed to make Ukrainian nationalists heroes; wound up with thousands of pieces of burned equipment, seen the sinking of the flagship of the Black Sea Fleet; lost an unknown number of soldiers both wounded and killed; seen an increase many times over of Russia's dependence on China; seen an acceleration in the decline of the standard of living; marked the resurrection of the Iron Curtain; and witnessed the growing dilapidation of the economy. As a bonus, the main sectors of the large, destroyed city of Mariupol were captured and now needs to be restored at its own expense. The Russian Federation, as a contradiction, has essentially contributed to the arming of Ukraine by causing an increase in its military budget of more than 10 times. It made Zelensky a popular world leader who now speaks in parliaments the world over. It has accelerated the entry of Ukraine into European institutions. It has managed to frighten and draw neutral Sweden and Finland into NATO. It has strengthened US hegemony in the world. It has made the Germans breathe a sigh of relief and calm down. In addition, the hatred between the peoples of Russia and Ukraine has been ignited to unprecedented degrees. If earlier there was a hatred among those on the margins, now there is a hatred for one another in all segments of the population in both countries.

---

38 Coynash, Halya. (2021, January 10) "Neo-Nazi Russian nationalist exposes how Russia's leaders sent them to Ukraine to kill Ukrainians." *Kharkiv Human Rights Protection Group, Information Portal "Human Rights in Ukraine."* "Dmitry Demushkin, a far-right Russian nationalist, has revealed details of how he was invited by the then Russian Deputy Prime Minister, Dmitry Rogozin, to gather nationalists to fight in Ukraine... [Rogozin] promised that, in return, Demushkin would be made 'mayor' of one of the cities in Donbas... In interviews, Demushkin mentions the effective torture methods used when the FSB began trying to 'convince' him to be cooperative. Demushkin refused and received a 2.5 year prison sentence which [began] with his being found guilty of organizing an extremist organization in March 2014... Unlike Vladimir Putin, who has no problem with claiming Russians and Ukrainians to be the same people, but using the first to kill the latter, Demushkin's refusal to recognize Ukrainians as a separate people logically demands that he is not prepared 'to kill his own.'" https://khpg.org/en/1608809502

17. In the course of the Special Operation, Russia is reminded of Suvorov's "science of winning."[39] It "demonstrates" this more by its obvious, flashy moves intended to prevent giving the impression that the RF is simply unable to deliver serious military actions to beat back the enemy. The military and political leadership of the Russian Federation, if it declares [that it is going to do] something, must act and not change the agenda along the way. Otherwise, no one will believe it, neither inside nor outside the country. "The state as a business project has completely failed," the nationalist patriots[40] maintain.

18. It is very important that the image positioning of those in power does not suffer. To do this, they need their decisions to be irreversible. An example of the opposite is evident when you overlay their various reactions to Finland and Sweden joining NATO: Recently it was announced that this *does not pose a direct threat* to Russia. But on July 1, 2016, Vladimir Putin assessed the likelihood of Finland joining NATO in *a completely different way*, stating,

"Imagine that Finland joins NATO. This means that the Finnish troops will no longer be independent, will no longer be sovereign in the full sense of the word. They will become part of the military infrastructure of NATO, which will suddenly appear on the borders of the Russian Federation." And Russia has changed its position on the issue of Ukraine's accession to the EU. First what was said was that this issue "lies on different plane" than membership in NATO. Now they say, it should not be allowed, as was announced about Ukraine's interest in joining the North Atlantic Alliance.

19. Of course, in a rapidly changing environment it is difficult to predict for the *long* term. A new combination of circumstances sometimes creates an unforeseen reality. But in the *short* term, you cannot miscalculate. The image of the powers that be suffers when they show a "lapse of memory." (Does the leader have sclerosis? Is it that the General Staff is so lousy? Is it that seven Fridays a week are spent in decision-making?).[41]

---

39 Aleksandr Vasilievich Suvorov is considered one of the greatest military commanders in Russian history. He was the author of several military manuals, the most famous being "The Science of Victory" [Russian "Наука побеждать."] He never lost a single battle he commanded. https://en.wikipedia.org/wiki/Alexander_Suvorov

40 *National patriots* [Russian "национал-патриоты"] refers to those adhering to the ideological and political movement of the late 1980s-1990s, which advocated the preservation of national historical cultural values and traditions of Russia on the basis of extreme nationalism. https://ru-wiktionary-org.translate.goog/wiki/национал-патриотизм?_x_tr_sl=ru&_x_tr_tl=en&_x_tr_hl=en&_x_tr_pto=sc

41 "Seven days a week are spent in decision-making" [Russian "Семь пятниц на неделе в принятии решений"] is a Russian idiom that means someone is utterly unreliable, fickle, and never follows

20. The philosopher Friedrich Nietzsche considered war to be the "means of cleansing" peoples and a primary way of regulating values. He believed that war should be taken not as a challenge, but as a given, taken the way the ancient Greeks did according to their military policy. In its crucible, the "debris"—the obsolete or false—of the old world is burned and is smelted to yield an outline for a new world order. A large-scale "disjunction" is taking place in the sense of achieving either a victory of the imperial-Russian project[42] or the triumph of a "new Ukrainian" national creation. Or maybe, as a result of this existential challenge of fire and blood, the paths of both the "Russkies" and the "Ukry" will merge, and in a single liberating rush, return to both peoples their true national identity so that they can then dispose of it themselves and live as one people or "independently" in a civilized way—sovereign, but in eternal peace. Interestingly, the head of European diplomacy, Josep Borrell,[43] said: "This war [the Special Operation] must be won on the battlefield."

21. The conflict unleashed by covertly bargaining for the exchange of "captured Nazis" is beneficial for the corporate interests of those involved in the war. The State Duma "is studying the possibility of exchanging captured Azov commanders for Ukrainian party functionary Viktor Medvedchuk,"[44] Zelensky's desired replacement and Putin's "godfather." This development has changed the focus of the rhetoric in the Russian Federation: Now the Ukrainian army is no longer made up of "Ukronazis," but of "Russian soldiers led by fascist officers and American generals," writes one of the Kremlin

through on what he has promised.

42 (2022, June 17) "Putin and the Project of a Big Russian Nation." *Cosmonaut Magazine*. This article provides a comprehensive discussion of President Putin's expansionist ideas vis-à-vis the Special Military Operation. https://cosmonautmag.com/2022/06/putin-and-the-project-of-a-big-russian-nation/ See also Hartnett, Lynne. (2022, March 2) "The long history of Russian imperialism shaping Putin's war." *The Washington Post*. In the article, the author provides a short but detailed history of Russia and its imperial past that has led to Russian military operations against Ukraine. She writes, "Reviving the imperialist dreams of his czarist forebears, Putin [has] moved to reclaim the empire that he believes was unjustly pilfered from Russia." https://www.washingtonpost.com/outlook/2022/03/02/long-history-russian-imperialism-shaping-putins-war/

43 Josep Borrell Fontelles is a Spanish politician serving as High Representative of the Union for Foreign Affairs and Security Policy.

44 Victor Volodymyrovich Medvedchuk is a Ukrainian lawyer, business oligarch, and politician who served as the chairman of the pro-Russian political organization Ukrainian Choice from 2018 to 2022. President Putin is godfather to Medvedchuk's youngest daughter. In May 2021, Medvedchuk was accused of treason and of attempted looting of national resources in Crimea and was put under house arrest. He escaped this house arrest on 28 February 2022, four days after the 2022 Russian invasion of Ukraine. In April, he was arrested by the Security Service of Ukraine. https://en.wikipedia.org/wiki/Viktor_Medvedchuk

ideologists, Sergei Markov.

22. The emphasis on the "terrorist nature" of the Ukrainian resistance to "aggression"—a blurry mix of "militants," "Nazis," and "nationalists"— is beneficial to many. If they are "terrorists," then the Geneva Convention on prisoners of war does not apply to them. They are under the protection of the Investigative Committee and the Prosecutor's Office. And since it is not formally a war, but, rather, an *anti-terrorist special* operation, then it is being made to fit into the classification of a terrorist organization like ISIS. Of course there are servicemen in the Ukrainian army who hold the views of ultra-right radicals, but their cry "Kill the Russian!" is not necessarily a cry for committing the genocidal atrocity associated with "fascists." What I am saying here is *not in defense of sadistic fanatics*, of those wounded in battle, of those who shoot captives made to get on their knees or those who rape their victims, but to avoid making one broad false accusation against everyone. In fairness, let me remind you that during the Patriotic War, the writer Ilya Ehrenburg called for "killing a *German*" in the article "Kill!" (published in the newspaper "Krasnaya Zvezda" of July 24, 1942): "If you cannot kill a German with a bullet, kill a German with a bayonet. If there is calm in your area, if you are waiting for a fight, kill a German before it starts. If you let a German live, he will hang a Russian man and dishonor a Russian woman. If you have killed one German, kill another...".[45] [Act] in war as [if you are] in war.[46]

23. However, what kind of symbiosis is this: possessing a "pro-Nazi worldview" yet also being drawn to "European norms"? For TV journalist Vladimir Solovyov, these things are organically compatible, even negatively synonymous: "Modern Europe has not broken away from the origins of Nazi doctrine." But what about Dostoevsky's "sacred stones of Europe" then?[47] I would propose

---

45 Ilya Grigoryevich Ehrenburg was a prolific writer and journalist, one of the most effective Soviet spokesmen to the Western world. When the "Thaw" began to reverse, he was censured for including material not considered proper for Soviet authors. Nonetheless he survived and remained prominent in Soviet literary circles until his death in 1967. https://www.britannica.com/biography/Ilya-Grigoryevich-Ehrenburg

46 Interestingly, "In War as in War" is the title of a 1968 Soviet black-and-white full-length feature film about the everyday life in combat [circa 1943] of a self-propelled gun crew during the liberation of the Right-Bank Ukraine [from the Germans]. https://ru-m-wikipedia-org.translate.goog/wiki/На_войне_как_на_войне_(фильм)?_x_tr_sl=ru&_x_tr_tl=en&_x_tr_hl=en&_x_tr_pto=sc

47 The source of the phrase "Sacred stones of Europe" are novels by Fedor Mikhaylovich Dostoevsky, including "Teenager" (1875) [Russian "Подросток"] and "Brothers Karamazov (1879-80) [Russian "Братья Карамазовы"]. In "Teenager," the character Versilov says, "For the Russian, Europe is as precious as Russia; every stone in it is sweet and dear. Europe was our Fatherland just as surely as Russia...". https://dic-academic-ru.translate.goog/dic.nsf/dic_wingwords/3473/Священные?_x_

that it is not "God is dead!", as the philosopher Nietzsche confidently believed, but that God's entourage has faded away for those who have forgotten him. But the imminently holy nature of Europe lives on in those who need it. The world, according to the philosopher Arthur Schopenhauer,[48] is "will and representation." The question is: Was the great poet Vladimir Mayakovsky carrying out a "fascist" denunciation of himself when he wrote, "I love to watch how children die"? [49] So "don't compare; he who lives is beyond comparison."[50]

24. The phenomenology of this "centaur-like" merging of the light and dark worlds in man is complex and has not been studied. I believe there is a situational transformation,

forced, superficial, and with an outward appearance—a temporary anomaly and complete transformation into a fiend—an inveterate villain. In the view of Philosopher Vitaliy Darensky,[51] this *Ukrainian post-Soviet national "non-identification" is a failed-nation syndrome.* The motive is "vindictiveness" ("ressentiment").[52]

25. Let me explain. In contrast to "non-identity," identity is a stable platform of values, beliefs, and convictions that determine both the daily behavior of people and their long-term life programs. Thus having ripped apart the Russian-Soviet root system that Ukraine shared with Russia, and in so doing, departing from the true identity of the

---

tr_sl=ru&_x_tr_tl=en&_x_tr_hl=en&_x_tr_pto=sc

48 "Arthur Schopenhauer." (2021, September 9) *Stanford Encyclopedia of Philosophy.* The 19th century philosopher, Arthur Schopenhauer, "asserts that among all the objects in the universe, there is only one object relative to each of us—namely our physical body—that is given in two entirely different ways. It is given as representation (i.e., objectively; externally) and as will (i.e., subjectively; internally). "The World as Will and Representation [German "Die Welt als Wille und Vorstellung"], completed and published in 1818, was his most famous work. https://plato.stanford.edu/entries/schopenhauer/#4

49 This line is from Vladimir Mayakovsky's poem "A few words about myself" [Russian "Несколько слов обо мне самом"] and is a classic example of his early satire. https://v--v--mayakovsky-ru. translate.goog/books/item/f00/s00/z0000001/st017.shtml?_x_tr_sch=http&_x_tr_sl=ru&_x_tr_tl=en&_x_tr_hl=en&_x_tr_pto=sc

50 This is the first line and title of Osip Mandelshtam's poem "Do not compare: he who lives is incomparable..." ["Не сравнивай: живущий несравним... "] written on January 18, 1937. https://45parallel-net.translate.goog/osip_mandelshtam/ne_sravnivay_zhivuschiy_nesravnim.html?_x_tr_sl=ru&_x_tr_tl=en&_x_tr_hl=en&_x_tr_pto=sc

51 Vitaly Darensky is a professor at Lugansk State Pedagogical University. https://luguniv.academia.edu/VitaliyDarenskiy

52 *Ressentiment* is a philosophical/psychological term coined by 19th century philosopher Friedrich Nietzsche. It refers to a psychological state arising from suppressed feelings of envy and hatred that cannot be acted upon, frequently resulting in some form of self-abasement. Ressentiment is not to be considered interchangeable with the normal English word 'resentment.' https://psychology.fandom.com/wiki/Ressentiment

Soviet-wide "community," including the identity of the Ukrainian ethnicity as it pertains to the great USSR-Russian history, the Ukrainians are trying to build a pseudo-identity from the wreckage of tribal consciousness, on the one hand, and from a superficial, epigonic assimilation of elements of "Europeanism" on the other. As a result of the utopian nature of these attempts, its identity is being simulated while, in reality, *there is no identity at all*. And this is precisely the reason for the aggressive Russophobia of modern "Ukrainianism." Behind this Russophobia is "an unconscious sense of guilt for betraying the real—not fictional—community created by Russian history," the history of a "trinitarian" people (Russians, Ukrainians, and Belarusians). This period of self-determination exposed the total degradation of the "neo-ukry"[53] (as, indeed, of the "new Russians"). That is, the "nazification" is retribution for breaking the umbilical cord of a common socio-genetic blood destiny with Russia. I would call this ethical and ideological anomaly a lost-identity syndrome and an attempt to replace that lost identity by coopting someone else's. The term "Nazism" (the wording having been established as a precedent by the verdict of the International Military Tribunal in Nuremberg dated October 1, 1946) is used to characterize Ukrainian radicals in Russia due to their behaviorally aberrant resemblance to the "German Nazis."

26. Darensky's concept is interesting in terms of clarifying the possible impact of "*non*-identification" or "*pseudo*-identification" (using a false alter ego) on consciousness and behavior. Simulating an identity has a compensating aspect (right up to point of the identity becoming complete). At the same time, it is hardly legitimate to base this phenomenon on the "*real absence*" of identification as an established fact. I assume that [the identity] remains, but in a distorted form, like a collage made by overlaying new layers that continuously change the image. In doing so, Nazi-style attributes are viewed as the dominant criteria reflecting the moral make-up of the Ukrainian resistance. According to the philosopher John Locke, the identity of a person consists of the totality of his uninterrupted consciousness and not of any substance (regardless of whether it is material or not)—"one and the same person is formed *not by the same substance*," but by the *same continuous consciousness*. According to the psycho-social understanding of self-identity, identity is the *maintenance of a balance between internal integrity and the values of society*. Thinker Mikhail Bakhtin believed that the *real "I" is always observed at the points where there is a discrepancy between a person and himself*, that is, in his iden-

53 The term used here to refer to Ukrainians—"ukry"—is pejorative.

194

tification with the "Other."[54] Consequently, the new self-identification could have grown into an equivalent "other I," complete in the sense of the completeness of its formation of the Alter Ego. That is, appearing to be a mutant, it is completely whole in the mirror reflection of its "I." The original archetypes of a something vaguely different lie dormant in it, but its essence—updated and reworked beyond recognition—is completely realized: The "Banderism" syndrome was genetically lying dormant a priori. It went through a baptism of fire and blood in World War II; it suffered years of persecution (jail, exile); it cooled down during Khrushchev's "Thaw"; it put out roots and made an appearance in post-Belovezha Russia; and recently materialized in a clash with the occupiers of independent Ukraine—its second "I".

27. And indeed "identity politics" itself makes it possible for a person to influence information about himself that becomes part of society. And finally, it is fashionable in certain strata to have "cool" tattoos. Therefore, adjacent ethnic groups of one people form a unity of opposites—"my brother is my enemy." "Ukraine has always been bipolar (orange-blue)[55]; in Ukraine, reprisals are visible and public. Russia is a more subtle and global system, connected both with the conditional global Kissinger and the conditional global Soros more closely than is Ukraine," writes M. Shevchenko. "Neither nickname, nor religion, nor the blood of ancestors itself makes a person belong to one or another nationality. The spirit, the soul of a person—that is where you need to look for determining whether it belongs to one or another people. How can you determine where the spirit belongs? By a manifestation of the spirit—a thought—of course. Whoever thinks in what language belongs to that people. I think in

---

54 Demidova, Elena Viktorovna. "The appearance of the Other in early M. M. Bakhtin. [Russian "Появление Другого у раннего М. М. Бахтина"] *Cyberleninka On Line Library.* "In an early essay, 'On the Philosophy of the Action'...the 'Other' is not presented as a person; it is one of the objects surrounding the Self. An analysis of Bakhtin's early works makes it possible to trace the process of the birth of manifestation of the Other for the Self... Bakhtin identifies the following structural components: two participants, I and the Other as two value centers, and three "worlds"—space, time, and meaning." https://cyberleninka.ru/article/n/poyavlenie-drugogo-u-rannego-m-m-bahtina

55 "Ukraine's bitter west-east schism is reflected in the political deadlock between its 'Orange' and Blue parties that has nearly paralyzed the state for the past year... the two sides remain separated by language, religious traditions, societal histories, and geopolitical preferences... The western part of Ukraine...was part of the Catholic states of Austria-Hungary and Poland for hundreds of years before Soviet dictator Joseph Stalin forcibly annexed it to the Soviet Union after World War II... While many...in the Ukrainian-speaking, nationalist west think the anti-Soviet veterans should be given military pensions and treated as Ukrainian patriots, their demands provoke fury in the heavily Russified east of Ukraine, where most accepted Soviet rule and millions served in the Red Army." (2007, September 28) "Ukraine's orange-blue divide." *The Christian Science Monitor.* https://www.csmonitor.com/2007/0928/p04s02-woeu.html

Russian," wrote Vladimir Dal (born at the Lugansk mining plant, in the current Donbass).[56] *So I offer a new element within the field of social science—a special military operation as a means of correcting two anomalies—the "Bandera" self-identification of Ukraine and the Russian state as a state corporation. Getting rid of the "new-style Russianness" as well as "Ukronazism" for the sake of returning them to their original ethno-images, is the future of a single people of three hypostases.*

28. Hiding behind hostages is not only a tactic of terrorists, but also a strategy of long-standing battles. And indeed, in itself, a "human shield," after all, can be formed on a voluntary basis, an act of self-sacrifice to support military undertakings. Responding to the call of patriotic consciousness—to help, as much as possible, with the Fatherland in danger, and "not sparing life or limb." In the world of topsy-turvy moral and ethical values, legalism, and formalized, fragmented consciousness, the unconventional is immediately

labeled "terrorism"—"ISIS"—"Nazism." But without trial or investigation, it is impossible to classify this elite of the Ukrainian resistance as "Ukronazis" in a single stroke, and then, just as unmotivated, to whitewash everyone at once, saying that only the foreign bosses of the troops are scum and filth.

29. What is needed in the Russian Federation, in the Ukraine, is the "decontamination" and "disinfecting" of a sick society. War is an indicator, a "litmus test" of civil usefulness and a stimulator of the healthy spirit of a nation. The gravestone equalizes *all* combatants, defenders of their homeland. And the ultra-patriots—denounced in the Russian Federation as "nationalists" and likened to "Nazis"—are often real zealous patriots, for the most part fearless warriors, "kvas patriots."[57] I'm not talking about sadistic torturers, pathological perverts. It is possible that the "hot-blooded" on both sides—the nationalist-"ukry" and the Avvakum-type[58] frenzied "Russians" (horribly nicknamed the

---

56 Vladimir Ivanovich Dal was a 19th century Russian language lexicographer, polyglot, Turkologist, and founding member of the Russian Geographical Society. https://en.wikipedia.org/wiki/Vladimir_Dal#Legacy

57 A "kvas" patriot [Russian "квасной патриот"] is an ironic expression signifying an individual's unconditional, ostentatious, blind, stubborn patriotism. Originating in the 19th century, it was used to refer to someone who always has praise for anything related to the Fatherland and condemnation of everything alien. https://dic.academic.ru/dic.nsf/ushakov/833034

58 Avvakum (1620-1682) was a Russian Orthodox archpriest who fought against the liturgical reforms of Patriarch Nikon. He is usually considered the principal leader of the Old Believers and was considered a religious zealot. Ultimately he was excommunicated and exiled to a remote prison colony beyond the Arctic Circle. https://www.encyclopedia.com/places/africa/so-tom-and-principe-political-geography/avvakum

"red-browns")[59]—will face military actions to clean out the common home and recreate a single people. And indeed national humiliation has come to the point that Polish Prime Minister Mateusz Morawiecki dared to openly declare the need to *destroy the "Russian world."* Victory in the Special Operation would strengthen the Russian world many times over, while its defeat would drive the pan-Slavic imperial resurrection underground. This is an existential operation—to exist or to not live. "Why do we need such a world if there is no Russia there?" Putin formulated the key phrase of Russian self-consciousness.[60]

30. Without nourishing the soil, there will be no inspiring victorious spirit. However, the umbilical cord binding the troops to the soil has been destroyed in Russia just like the now boarded-up windows of the villages. There is a threat of turning into "Ivans, not remembering kinship"[61] or into "mankurts" (a word

from Chingiz Aitmatov's novel "The Buranny Railway Stop")[62]—individuals who have completely lost touch with their historical, national roots, forgetting about their kinship. This mentality is what forms a sense of the Motherland and is the foundation of a national identity.

31. Vladimir Putin is a black-belt in judo, truly a superman of tradition in an anomalous world of values turned upside down. If his decision-making is indeed "sclerotic," it does not mean that he is in the least senile or suffering from dementia, that he is becoming "brain dead," or that he has gone crazy over either communism or Nazism. That he is clinically abnormal. That he is behaviorally "sadomasochistic" (in levying reciprocal sanctions). As a judoka, he waits for his "partner's" mistake to throw him to the mat. The West, forcing Russia, if not to transform, then to "self-destruct," is committing hara-kiri like a samurai: It is pushing Russia towards

---

59 "Red-brown" is a propaganda cliché implying the union of communist and far-right (fascist, Nazi, etc.) forces and/or ideologies. https://dic-academic-ru.translate.goog/dic.nsf/ruwiki/649414?_x_tr _sl=ru&_x_tr_tl=en&_x_tr_hl=en&_x_tr_pto=sc

60 See (2018, March 7) "'Why would we want a world without Russia?' Putin on Moscow's nuclear doctrine." *RT.* https://www.rt.com/news/420715-putin-world-russia-nuclear/

61 "Ivan, who does not remember kinship" [Russian "Иван, не помнящий родства"] is an expression...that refers to a person who does not remember and does not observe traditions, who does not respect the customs of his ancestors, and who has renounced his environment. https://esperan to--plus-ru.translate.goog/fraz/i/ivan-ne-pomnjashcij.htm?_x_tr_sch=http&_x_tr_sl=ru&_x_tr_ tl=en&_x_tr_hl=en&_x_tr_pto=sc

62 In Chingiz Aitmatov's novel "The Day Lasts More Than a Hundred Years" [Russian "И дольше века длиться день"], Aitmatov addresses the legend of the mankurt which describes how to create ideal slaves by erasing their memories. The novel provides a biting allegory for the loss of Central Asian identity and language during the Soviet era. The novel was printed in a magazine version under the title "The Buranny Railway Stop" [Russian "Буранный полустанок"] 1980 in *Novyi Mir #11.* https://sites.pitt.edu/~filmst/events/TurkmenFilmSeries/mankurt.htm

a strategic partnership with China, even risking turning the United States into the Disunited States of America... And the "Anaconda Strategy"[63] is not letting up: it is in the midst of a third suffocating phase against the Russian Federation—the phase to bring about "the liquidation of the Russian Federation within its current borders, with a complete loss of economic, political, and military sovereignty."

32. For all that, the trump cards have been trumped or at least their reputation has been tarnished: the favorable moment for a coup in Russia, has, if you will, has passed. (In Belovezha there was, I believe, not enough of a coup—a falling short of the goalpost, if you will.-E.V.) The opposition has been beheaded. Mikhail Khodorkovsky[64] is in exile.

Alexei Navalny is in prison. Anatoly Chubais, with his project for a liberal-revenge dictatorship, has left the Russian Federation's political and ideological scene. Diplomacy has bitten the dust. The weapons of the apocalypse are the Poseidon torpedo[65] and the RS-28 "Sarmat" ICBM that are in readiness. And this ten-ton massive rocket launcher can deliver a deadly cargo even across the South Pole (after all, it is almost not controlled) and is capable, as the head of Roscosmos Dmitry Rogozin[66] authoritatively stated, to destroy half of the coast of a continent. And showing the West the craters from the "Sarmat" missile, in his view, would serve as the best recommendation for the West to be more polite when dealing with Russia.

33. Thus, a way of looking at the

---

63 "Today, the West is actively using the arsenal of hybrid war against Russia. Not deciding on a direct military confrontation, the United States and its allies are trying to strangle Russia by other methods, to tighten the Anaconda Loop on its neck. That is how the American Rear Admiral Alfred Mahan called his concept back in 1890." See (2018, April 24) "Will Russia break the 'Anaconda loop'?" *Reporter*. https://en.topcor.ru/1057-razorvet-li-rossiya-petlyu-anakondy.html.

64 Dawkins, David. (2020, 14 March) "Putin And Khodorkovsky Trade Blows As Presidential Power Grab Gathers Momentum." *Forbes*. "In 2003, Khodorkovsky argued with Putin at a televised meeting...implying that major government officials were accepting millions in bribes. Unsurprisingly... he was arrested for tax evasion, embezzlement and fraud...and was found guilty...and jailed for eight years." https://www.forbes.com/sites/daviddawkins/2020/03/14/putin-and-khodorkovsky-trade-blows-as-presidential-power-grab-gathers-momentum/?sh=1101621d2b54.

65   Paul, Jacob. (2022, April 21) "Putin's horror Poseidon 'super-weapon' capable of decimating entire cities." *Express UK*. "Poseidon is the Russian President's nuclear-armed submarine drone that can zip through the waters at a speed of 125mph after getting dropped onto the seabed. It is officially known as an 'Intercontinental Nuclear-Powered Nuclear-Armed Autonomous Torpedo'... [On 20 April], Russia unveiled a terrifying ICBM dubbed the "Satan 2" by NATO. Also called the RS-28 Sarmat, it can carry a 10-ton payload—up to 10 large warheads and 16 smaller ones." https://www.express.co.uk/news/science/1599255/putin-super-weapon-poseidon-drone-satan-2-sarmat-russia

66   Balachuk, Iryna. (2022, July 12) "Director of Roskosmos Dmitri Rogozin posts picture of Sarmat nuclear missile, again says it will be serially produced." *Ukrayinska Pravda*. https://www.yahoo.com/video/director-roskosmos-rogozin-posts-picture-071736570.html

dislocation of forces in the undeclared war between the West and Russia is as follows: Washington (the EU and Zelensky) vs. Russia (alone, with a "fifth column" in its rear, with some "asymmetry" in decision-making and incomprehensibility of the reasoning behind it). In Ukraine, in fact, there is a Civil War where shackled to one single chain are Slavs related by blood: traditionalists (like Soviet miners under the banner of a common victory in the Second World War) and neophytes (who have been remade on a broad scale into a self-identified pro-Bandera demographic). The Red "terrorists" and the White Guard torturers fought just as violently and furiously. But let's separate the "cutlets from the flies": In addition to monsters by nature (scoundrels by birth is a phrase from somewhere in the early Dostoevsky writings), there may be an ultra-nationalist element, which was formed as a result of the prevailing negative circumstances. An example of this is the fact that a manifestation of national dignity, long restrained by the authorities and designated by them as illegal, can break free "with great force," like a genie from a bottle, and take revenge on the world for its painful past. Intellectuals who have undergone moral abuse, moral torture, often bear the appearance of individuals bitterly "laughing through their tears"—possessing an avenging "scarred" consciousness syndrome.[67]

34. In the "dashing 90s," true patriots of Russia turned out to be superfluous in their Fatherland (including your humble servant). They were unnecessary to the anti-national regime of "criminal revolutionaries" established in the Russian Federation. "Aborted" by Russia—expelled abroad—they turned out to be in demand, in particular, by the anti-Russian oligarchy of "Ukraine." And to this day, the "Nazis"—the stronghold of the inveterate oligarchic Kyiv regime—serve as guards ensuring "not a step back!".[68]

35. The moment to put the squeeze on Russia has passed. The recent intensification of Russian military efforts vis-a-vis Ukraine prompted the most experienced diplomat and former US Secretary of State Henry Kissinger to urgently recommend a speedy peace: "The West should give up trying to achieve a military defeat for Russia in Ukraine, and the latter should make territorial concessions." Apparently, in order to preventively stall the consolidation of Russia's growing success. And so that only the current limited

---

67 I delved into these studies long ago, when I was a dissertation candidate exploring the topic "The problem of ambition in Dostoevsky as a factor in Russian consciousness in decision-making." - E.V.

68 On July 28, 1942, Stalin issued Order No. 227, what came to be known as the "Not one step backward" [Russian "Ни шагу назад!"] order, in light of German advances into Russian territory. The order declared "Panic-makers and cowards must be liquidated on the spot." https://en.wikipedia.org/wiki/Order_No._227

territories recaptured from Ukraine would remain under Russian Federation control (just as the troops of Zhukov and Patton hurried towards each other and as a result, met on the Elbe). And only those territories. This is what, after a short pause, would allow a "Pyrrhic victory" to turn into the total defeat of Russia. By sacrificing a part, you save the whole (the logic of Lenin's "Brest—obscene—peace").[69]

36. But the political bureaucracy does not need anyone to have a speedy victory since it would inevitably entail a change of bankrupt "elites." Therefore, by sabotaging the exit strategy of having Ukraine sign an act of surrender, both sides will be supporting the scenario of a war "to a victorious end." Politburocracy "doesn't need a concrete military victory, but, rather, the eternal anticipation of one." Therefore, the "global hybrid war" will drag on "until the last Ukrainian" or until "the last dollar in the US treasury." Even for the most experienced general like Vladimir Shamanov, the question of the future of Ukraine is difficult to predict. One can only assume that the demilitarization of the country may take from five to ten years, and it will be difficult

to form a government there that is "not tarnished by neo-Nazis." At the same time, Shamanov believes that if these two phenomena do not occur, Ukraine will not be suitable for Russia as a new subject in its federation.

37. But it is too early to "share the pelt of a bear that hasn't been killed yet" and it's ridiculous if we are talking about a blood relative. But for the sake of having a single tri-hypostatic Russian people then it would be a welcome development. Patriot Dmitry Rogozin has just proposed to "improve life" at Ukraine's space factories. When the world was bipolar—American-Russian—global security was maintained with limits that were clear to all: neither war nor peace. Shouldn't we not move in this direction? Otherwise, the challenge of China will relegate to oblivion the civilized conflict between the United States and the Russian Federation.

38. The highest level of the *nomenklatura* of both Russia and Ukraine is trying with all its might to thwart this ongoing "conceptually cleansing war" (the SMO—Special Military Operation). Stopping the advance of the troops is pretty

---

69 The Treaty of Brest-Litovsk was a separate peace treaty signed on March 3, 1918, between the new Bolshevik government of Russia and the Central Powers (German Empire, Austria-Hungary, Bulgaria, and the Ottoman Empire), that ended Russia's participation in World War I. The treaty terms included Russia's recognition of the independence of Ukraine, Georgia, and Finland; the Russian surrender of Poland and the Baltic states of Lithuania, Latvia, and Estonia to Germany and Austria-Hungary; and the ceding by Russia of Kars, Ardahan, and Batum to Turkey. https://en.wikipedia.org/wiki/Treaty_of_Brest-Litovsk

much a consequence of the agreement[70] made with the "Banderites." Russian Defense Minister S. Shoigu explains the slowdown in the advance of Russian troops in Ukraine as merely a strategic maneuver that guarantees the minimization of losses among the Ukrainian civilian population. Such tactics are contrary to the basics of military art. For the truth is indisputable: High offensive rates lead to a significant reduction in the losses of the attackers and to high losses of the defenders. Has the General Staff really forgotten this axiom of the "science of winning," knowing by heart the monograph of the theorist V. Savkin "The Pace of Advance"?[71] Or is it that the military is carrying out—within allowed limits—exactly the policy that has been proscribed to them as a result of the secret collusion between the ruling elites of both countries? After all, in the military manual used by the generals it clearly states, "With a high rate of advance and effective suppression of enemy fire, the

attacker suffers fewer losses in manpower and equipment. With an increase in the rate of advance, there is a further decrease in losses." In a half year of battle, the Russian army has deliberately been restrained—prevented from crossing the borders into Pridnestrovie (Transnistria),[72] hoisting the flag over Odessa. Ruling regimes like these have, on purpose or by accident, increased the losses on both sides (some with business-like pragmatism, others in a mindless rampage, fighting "to the last Ukrainian"—even if he is poorly equipped). They should be replaced by the nationally sane (rule by moderate nationalists). The war will make more clear national perspectives and could grow into civil wars in both countries—"for our freedom and yours." And whether to fight alongside the supranational oligarchy is something to be decided by those peoples who are pitted one against the other. In this lies the existential meaning of the military and mental clash within the Eastern Slavic ethnos.

---

70   The slang form of the Russian word for "agreement" used here—[Russian "договорняк" rather than the neutral "договор"] is taken from sports jargon to refer to a competition where the participants have agreed in advance on the result. https://kartaslov-ru.translate.goog/значение-слова/договорняк?_x_tr_sl=ru&_x_tr_tl=en&_x_tr_hl=en&_x_tr_pto=op,sc

71   Colonel Vasiliy Yefisovich Savkin, who served on the faculty of the Frunze Military Academy, wrote "The Basic Principles of Operational Art and Tactics," which was published in 1972... It was probably the most important Soviet military publication of that year. As a candidate of Military Science, he wrote "How to Achieve a High Pace of Advance under Modern Combat Conditions" (1962) and "Rates of Advance" (1965). https://books.google.com/books?id=UxBIAQAAIAAJ&pg=PR5&lpg=PR5&dq=v.+savkin+%22pace+of+advance%22&source=bl&ots=v52iWM9QzP&sig=ACfU3U3TNoC5V3laN0_d8q2egAgazC3BRQ&hl=en&sa=X&ved=2ahUKEwio77SH0-L5AhXEFmIAHSqQBHAQ6AF6BAgCEAM#v=onepage&q=v.%20savkin%20%22pace%

72   For a discussion on Odessa and Transnistria vis-à-vis the Military Special Operation, see Ferris, Emily. (2022, April 11) "The Battle for Odessa and its Railways: Could Transnistria Assist?" *Royal United Services Institute (RUSI)*. https://rusi.org/explore-our-research/publications/commentary/battle-odessa-and-its-railways-could-transnistria-assist

Dr. Evgeny Aleksandrovich Vertlib / Dr. Eugene Alexander Vertlieb President of the International Institute for Strategic Assessments and Conflict Management (MISOUK-France); executive editor of the Western Policy Forecasting Department of Slavic Europe (Munich); executive member of the Lisbon-Vladivostok Initiative (Paris).

Dennis T. Faleris received a B.S. from the University of Michigan and a Master's degree in Russian Linguistics from Georgetown University. For more than thirty-five years, Mr. Faleris worked as an instructor, translator, senior intelligence analyst, and intelligence production manager at the National Security Agency. His career centered around Soviet/Russian military issues as well as a variety of transnational issues. He currently resides in Annapolis, Maryland, with his wife, Kathleen.

Когнитивно-стратегический смысл военной спецоперации на Украине

Д-р Евгений А. Вертлиб

1. «Советский коммунизм содержит семена своего собственного саморазрушения и в конечном итоге рухнет из-за внутренней слабости» - сказал именуемый «архитектором холодной войны» политик Джордж Кеннан. Две детонационные волны способствовали обрушению СССР. *Сперва* кремлёвским заговором «врачей-вредителей» ликвидировали «отца народов». В результате госпереворота, подстрахованного парой сот танков у Кремля и бомбардировщиками на взлёте, наступила «хрущёвская оттепель». Были амнистированы бывшие боевики ОУН-УПА, бандеровцы, власовцы и прочие коллаборационисты. Как сказано в недавно рассекреченной справке, «оуновцев стало в десять раз больше и напряжённость возросла в разы по всей Украине». Вот они истоки «майдана». Устои Красной империи зашатались. Новейшая история России переворотно-суицидальна.

2. *Затем* Беловежским заговором высших номенклатурщиков порешили сам Советский Союз. Прекратил существование - «самоубился». Хотя суицидальной наклонности у державы не наблюдалось. Даже наоборот: СССР вполне уверенно балансировал на плаву. Юрий Андропов силился адаптировать китайскую практику Дэн Сяопина со «свободными экономическими зонами» - опыт уживания коммунизма с технологической реформацией. И Михаил Горбачёв вполне уверенно клеймил «так называемых демократов», которые «готовят государственный переворот» и вынашивают коварные планы «раздробления нашего великого многонационального государства».

3. Но дела говорят громче слов. Госсистема совершала смертельные ошибки. Стоило в 1989-м сдать ГДР – посыпался весь соцлагерь. Падению «железного занавеса» способствовало изъятие из «брежневской» Конституции СССР статьи шестой: «о руководящей и направляющей роли КПСС в советском обществе». Эта демократическая коррекция основ государственного конструкта имела пагубные последствия для власть предержащих.

4. С извлечением из фундамента державного монолита

этого основополагающего «кирпичика» было нарушено отлаженное функционирование государственной машины: разбалансирован маховик управления; устои системы и каркас конструкции государства были обречены. Вскоре Борис Ельцин отрапортовал в Конгрессе США: «коммунистический идол, *который* сеял повсюду на земле социальную рознь, вражду и беспримерную жестокость, который наводил страх на человеческое сообщество, рухнул. Рухнул навсегда. *И я здесь для того, чтобы заверить вас: на нашей земле* мы не дадим ему воскреснуть!». Однако спецоперации «Наш Крым» («Вежливые люди») и нынешняя по демилитаризации и денацификации «укронацистов» - разворот тренда на имперские позывные России.

5. Заметим, что после поражения в холодной войне РФ проводила *инерционную* внешнюю политику. Только форс-мажор складывающихся обстоятельств побуждал высшее начальство «что-то предпринять». Так война 2008-го между Грузией и Россией произошла из-за нерешительности Кремля в присоединении Абхазии и Южной Осетии (убийство миротворцев – лишь casus belli). Неологизм *«Крымнаш»* (2014) - ответная реакция «мягкой си-

лой» на вытеснение российского флота из Чёрного моря. Затем «всё уснуло», пока новая угроза ни пробудила РФ в холодном поту. Оказывается: «укры» намылились в НАТО (а это четыре минуты ракетно-подлётного времени до Кремля!). И грянул бой - военная спецоперации по сути по восстановлению стату-са-кво: диспозиции НАТО на 1997-й – до продвижения альянса на Восток (для сравнения: зачистка сепаратистского «ичке-ризма» называлась «наведением конституционного порядка»).

6. Если бы не будили «русского медведя», Кремль вряд ли пошёл бы на спецоперации. Политик Александр Проханов конста-тирует: Вдадимир Путин «был коромыслом весов, на котором качались две чаши укладов - патриотического и либерального. Но с какого-то определённого момента эти весы разбалан-сировались: патриотический уклад вышел из-под контроля Путина, либеральный - тоже. Президент не сумел за время после Крыма совершить дол-гожданного рывка, развития, которое бы соединило два этих уклада. И каждый уклад пошёл своей стороной. Внутри каждого возникла путаница, сложна-я система распадов. Поэтому Путин не контролирует эти два огромных массива современной России..., которые развивают-ся своим собственным путём,

причём достаточно хаотично». В Украине же, наоборот, война выявила подготовленность общества к организованному сопротивлению «москалям» и консолидацию элит. Одной из ошибок российской армии стало ожидание, что её на Украине «будут встречать с цветами», но эта составляющая уже преодолена, заявил генерал-полковник экс-командующий ВДВ Владимир Шаманов. На Украине ещё с советских времён сохранилось много военных вузов, которые есть практически в каждом регионе. Это позволило стране сохранить так называемую школу, «основу возрождения армии».

7. С поражением Советского Союза на антикоммунистическом этапе перманентной холодной войны между Западом и Россией социум фрагментизировался – на постсоветском пространстве состоялся «парад суверенитетов». Ведь 72-я статья Конституции СССР допускала выход республик из его состава «вплоть до отделения». Вместо прагматичной трансформации по-китайски, Россия «самоопределилась» по-ленински - «головотяпски» (выражаясь языком сатирика Салтыкова-Щедрина) решив независиться *от самой себя* (в умопомрачении приняв свои Союзные республики за «колонии»). Отвергнув идеологию

«совка», «прорабы перестройки» взамен кинули в массы компенсирующий её отсутствие клич лавочников: «Обогащайтесь!».

8. Под этим французским девизом «Enrichissez-vous!» «РФ как госкорпорация» вернулась к *феодализму паханат-компрадорского* руления страной. Этим слоганом французского политика-либерала Франсуа Гизо в 1925 попользовался идеолог большевизма Николай Бухарин – чтоб оживлением элементов капитализма при НЭПе помочь выжить беднякам в крайней нужде. Увы, нищих в РФ всё не убывает: «до конца 2022 года более 20% граждан России окажутся за чертой бедности».

9. Чужеземный образчик общественно-политической трансформации - «Перестройка» заждилась на «знаках французской революции»: «Если искать корни нашей перестройки, то можно дойти и до французской революции, а потом до Коммуны» - подсказал сам «Горби» наличие чужих лекал в проекте реформирования госсистемы. В «перестройке» маячила *перестрелка*. Сперва «колоний» СССР между собой, а затем и генетико-цивилизационных братьев – «москалей» и «укров». Тектонику развития драмы в геополитической трагедии

СССР я почуял уже в 1980-х: в своей статье под названием «Генезис и пути политического сознания при Горбачёве», с подзаголовком: «"Перестройка" – эволюция или **пристрелка перед перестрелкой**» -опубликованной в известном парижском журнале на пяти языках писателя Владимира Максимова «Континент» (№63, апрель 1990, с.189-223). Суицидальность в некоторой степени обнаружил Путин в принимаемых Западом решениях: «экономическим аутодафе выстрелил себе в ногу либо в голову».

10. «Великой криминальной революцией» назвал режиссёр Станислав Говорухин ликвидацию СССР. Но нет худа без добра: зато от «Перестройки» повеяло «Парижской коммуной»: с ней обозначился выход на подлинное народовластие. Но русскому национальному самодвижению противостоит *«компрадорская стабильность»* - что погружает российский социум в когнитивный диссонанс. Судя хотя бы по схлёстке двух концептов «подлинности патриотизма», представленных главой Чечни генерал-лейтенантом Рамзаном Кадыровым и пресс-секретарём президента Дмитрием Песковым.

11. «Дружеская» ли перепалка между ними -- двух перфекционистов Русского мира, или уже конфронтационная фаза предбаррикадного размежевания социума? «Не приведи Бог видеть русский бунт - бессмысленный и беспощадный» - заклинал Пушкин. В нынешней ситуации – не «бессмысленный» бунт. И вот почему. Судя по реакции на другой инцидент (недавнее нападение на посла РФ в Варшаве), генерал Кадыров твёрд и решителен: «Мы настроены очень жёстко, и *мы просим президента заставить все государства, которые с нами не считаются, встать на колени…* Почему не принимают меры?». Вопрос поставлен ребром и в прояснении смысла спецоперации на украинском направлении в концепте национальных ценностей и проводимой военной политики. Так «Во имя чего сапог землю растаптывает скрипящ и груб? Кто над небом боёв – свобода? бог? Рубль?» - пророчески вопрошал поэт-трибун Владимир Маяковский.

12. А не готовят ли чиновники поражение России? – задаются вопросом украинский политик Олег Царёв и российский предприниматель Андрей Самохин. Ведь нынешняя «элита» категорически не готова к мобилизации «за Родину» и

ждёт «управляемого поражения» России. Специальная операция, явно перешедшая в фазу затяжных кровопролитных военных действий, вплоть до возможных прямых столкновений с НАТО, требует совершенно иного управления, экономики, сознания народа и его элиты, чем было доселе и продолжает оставаться сейчас. «Мобилизация» – это и собирание в один кулак воли и духа всех составных государства. Однако вместо этого пока наблюдаются странные двусмысленности со стороны власть предержащих и их обслуги. Огромная армия чиновников или вовсе открещивается от войны («и зачем нам эта война?»), или хранит глухое молчание. Они хотели бы «если *проиграть, то контролируемо*, например, отдать всё или часть осво-бождённых территорий. Отдать Крым, но не сразу, а, например, с отсрочкой исполнения лет на пятнадцать - двадцать пять. Поражение и санкции, конечно, скажутся на уровне жизни: уровень жизни в стране понизится, но это не сильно скажется на элитах». Отсюда *осторожность, переходящая в саботаж.* Внутри российского руководства есть значительное лобби, позволяющее проводить со страной «любые геопо-литические эксперименты». Хотя патриотизм имманентно

присущ русскому народу, но *мандат доверия нынешнему режиму не вечен.* Не грозит ли РФ повторение на новом этапе известной трагической формулы последнего русского Государя: «Кругом трусость, обман и предательство»? В российском истеблишменте наблюдает-ся картина именно *ползучего предательства.*

13. «В такой ситуации старые элиты интуитивно настроены на поражение. Им будет лучше, если мы проиграем. Причём *лучше прямо сейчас*, пока про-тивостояние не зашло слишком далеко. А чтобы не зашло слишком далеко, не привет-ствуется и *не стимулируется набор добровольцев в России, не наносятся удары по критической украинской инфраструктуре, не вводятся симметричные меры в ответ на экономические и финансовые санкции: нет ответных арестов активов и долги западным странам продолжают выплачиваться. Отказ от нанесения ударов по «центрам принятия решений».* Всё делает-ся, чтобы продемонстрировать противнику, что мы готовы платить и каяться, каяться и платить» - негодует непримири-мая оппозиция. Недоумения в народе множатся, а внятно, не-противоречиво растолковать их власть от своего лица или хотя бы через штатных телепропаган-

дистов не хочет или не может.

14. Тут *всю систему* менять надо. «В нашей стране, - пишет экономист Сергей Батчиков, - по итогам "либеральных рыночных реформ" создана и действует уникальная по мировым меркам модель экономики: нелегитимная с точки зрения общественного мнения; незаконная с точки зрения источников формирования прав собственности; крайне неэффективная и непроизводительная с точки зрения общественного воспроизводства; неконкурентоспособная на мировом и внутреннем рынке; системно управляемая и манипулируемая извне; социально несправедливая; крайне ресурсо- и энергоёмкая; примитивная по своей структуре; деиндустриализированная и дезинтегрированная». Спрашивается: можно ли мобилизовать такую систему на противостояние военному и цивилизационному вызову, брошенному России Западом? Патриот Александр Проханов резюмирует: ***собранное Путиным государство сгнило почти дотла***». Нужно начать «переформатирование» системы с самого верха. А включённый режим «вставай, страна огромная» кардинально поменяет ситуацию и в

обществе, и на фронте, и в экономике. Но прежде всего всем осознать: *если не победа, то уничтожение*.

15. У высшего начальства РФ есть возможности всеми имеющимися средствами отстаивать интересы страны, но нет воли на то. Так и ядерное оружие они вряд ли превентивно задействуют. Эксперт спецслужб Яков Кедми выявляет болевые точки в системе российского принятия решений и безопасности. Он уверен, что России при стратегических козырях и играть следует иначе: «Достаточно одного или полутора залпов многоцелевой подлодки с «Цирконами», и порядка 50–60 электростанций Британии через десять минут перестанут существовать. *И вся Великобритания вернётся в каменный век*». Израильский аналитик приветствует смену вех в принятии внешнеполитических решений РФ: от *компромисса* во что бы то ни стало – к *решению* вопроса. «Лавров и российская дипломатия поменяли тон»: отход от практики «*анастезиологической*» (не причинить бы никаких болезненных неудобств Западу). Серьёзным изъяном считает эксперт избыточную доброту и сострадание россиян там, «где делать этого

совершенно не нужно». С точки зрения Кедми, и ***военная стратегия встречного или оборонительного ядерного удара***, которая принята в Военной доктрине России, а также в основах государственной политики РФ в области ядерного сдерживания, - подлежит корректировке. У встречного ядерного удара есть недостаток: «*Встречный удар не останавливает ракеты с ядерными головками, которые летят в вашу сторону. Это предотвращает только одно - **превентивный удар***»: он лишает противника возможности уничтожить, или нанести непоправимый ущерб вашему государству, а встречный – только пугает. А если они не испугаются? «А если они полуидиоты, а если они неправильно подсчитали? И ракеты, которые они послали, они полетят. Первая задача любого государства это защитить себя. Уничтожают противника, чтобы защитить себя». Яков Кедми уверен, что идёт против России **настоящая война, не гибридная**, нацеленная на **полное *уничтожение России, как государства до 2025 года.*** И в этом графике США действуют, применяя все возможные средства уничтожения. Поэтому сегодня, чтобы победить, Россия, по мнению Кедми, должна поменять все свои концепции и стратегии прошлого года, и *приспособить их к новому периоду военного противостояния с Западом на период 2022-2023 годов.* Россия не имеет права не выиграть в этой борьбе.

16. Русская победа пробивается микрошагами. Остатки СССР уничтожают друг друга. Россия, по подсчёту либерального обозревателя Дмитрия Демушкина, за три месяца спецоперации: потеряла золотовалютные резервы; собрала все возможные санкции; вылетела из мировых институтов; умудрились украинских националистов сделать героями; получила тысячи единиц сожжённой техники, утонувший флагман Черноморского флота, неизвестное число потерянных солдат убитыми и ранеными; увеличилась в разы зависимость России от Китая; ускорилось падение уровня жизни; воскрес железный занавес; полуразрушенная экономика... Как бонус захвачен фундамент крупного разрушенного города Мариуполя, который теперь надо восстановить за свой счёт. РФ методом от противного способствовала вооружению Украины, увеличив военный бюджет более чем в 10 раз. Сделала Зеленского популярным мировым лидером выступающим в парламентах

мира. Ускорила вхождение Украины в европейские институты. Умудрилась напугав, втянуть в НАТО нейтральную Швецию и Финляндию. Укрепила гегемонию США в мире. Заставила выдохнуть и успокоиться немцев. Разожглась до невиданных высот ненависть между жителями России и Украины. Если раньше это был удел маргиналов, то сейчас это ненависть всех слоёв населения обоих стран к друг другу.

17. Россия по ходу спецоперации вспоминает суворовскую «науку побеждать». Больше «демонстрируются» ею очевидные яркие шаги – чтоб не складывалось впечатления, что РФ просто не способна на серьёзный отпор. Военно-политическому руководству РФ, если оно что-то заявляет, надо действовать. И не менять по ходу повестку. Иначе никто ему не будет верить - ни внутри, ни во вне. «Государство как бизнес-проект полностью провалился» - утверждают национал-патриоты.

18. Очень важно чтоб не страдало *имиджевое позиционирование* власти. Для этого надо быть бесповоротными в принятых решениях. Пример обратного – накладка в толковании намеченного факта вхождения Финляндии и Швеции в НАТО: ныне объявлено, что этот акт

«непосредственной угрозы не создаёт» России. А ведь Владимир Путин 1 июля 2016 оценивал вероятие вступления Финляндии в НАТО *совершенно иначе*: «Представьте себе, что Финляндия вступит в НАТО. Это значит, что финские войска уже перестанут быть неза-висимыми, перестанут быть суверенными в полном смысле этого слова. Они станут ча-стью военной инфраструктуры НАТО, которые в одночасье окажутся на границах Рос-сийской Федерации». Россия изменила позицию и по вопросу вступления Украины в ЕС. Сперва: что эта тема мол «лежит в другой плоскости» нежели членство в НАТО. Теперь – не пускать, как и в Северо-Атлантический альянс.

19. Конечно, при быстро меняющейся конъюнктуре трудно прогнозировать на *долгосрочную* перспективу. Новая парадигма стечения обстоятельств образует иногда непредвиденную реальность. Но на краткосрочную перспективу не должно быть недоучётов. От «аберрации памяти» страдает *имиджевое позиционирование* власти (склероз у лидера? Плох Генштаб? Семь пятниц на неделе в принятии решений)?

20. Философ Фридрих Ницще войну считал «средством гигиены» народов и главным

ценностным регулятором. Он полагал, что войну следует принимать не как вызов, а как данность, принимать её по образу военной политики древних греков. В её горниле сгорает «рухлядь» (отжившее или ложное) старого мира и выплавляется очертание нового миропорядка. Происходит масштабная «дизъюнкция» - в смысле победы или имперско-русского проекта, или торжества «новоукраинского» национального созидания. А может быть, в результате этого экзистенциального испытания огнём и кровью сольются пути и «москалей» , и «укров» – в едином освободительном порыве вернуть обоим народам их подлинную национальную самоидентификацию, чтобы затем они распорядились ею сами – жить одним народом, или же цивилизованно «самостийно» - суверенно, но в вечном мире. Кстати, глава европейской дипломатии Жозеп Боррель так и заявил: «Эта война должна быть выиграна на поле боя».

21. И развёрнутая для скрытного торга коллизия с «пленёнными нацистами» выгодна корпоративным интересам фигурантов войны. В ГосДуме «изучается вопрос о возможности обмена захваченных в плен командиров "Азова" на Виктора Медведчука» (украинского

партийного функционера и кума Путина) – желанного сменщика Зеленского. И под это дело меняется акцентировка риторики в РФ: армия Украины - это уже не «укронацисты», но «русские солдаты, которыми руководят фашистские офицеры и американские генералы» - пишет один из идеологов Кремля Сергей Марков.

22. Акцент на «террористичности» (мутные «боевики», «нацисты» и «националисты») украинского сопротивления «агрессии» выгоден многим: если «террористы», то к ним не применима Женевская конвенция о военнопленных – они подопечные Следственного комитета и Прокуратуры. А раз формально не война, а *антитеррористическая спец*операция, то и подгоняется она под террористическую организацию «ИГИЛ». Конечно, в украинской армии есть военнослужащие и со взглядами ультраправых радикалов. Но их клич «убей русского!» - это не обязательно геноцидное злодеяние «фашистов». Говорю это *не в защиту садистов-изуверов*, в бою раненных, по коленям стреляющим и насилующим свои жертвы, а чтоб напраслину не возводить *на всех*. Справедливости ради напомню, что в Отечественную войну и писатель Илья Эрен-бург призывал «убить *немца* - в

статье «Убей!» (газета «Красная звезда», 24 июля 1942): «Если ты не можешь убить немца пулей, **убей немца** штыком. Если на твоём участке затишье, если ты ждёшь боя, **убей немца** до боя. Если ты оставишь немца жить, немец повесит русского человека и опозорит русскую женщину. Если ты убил одного немца, убей другого...». На войне как на войне.

23. Однако что за симбиоз: «пронацистское мировоззрение» и тяга к «европейским стандартам»? Для тележурналиста Владимира Соловьёва эти вещи органично совместимы, даже негативно синонимичны: «Современная Европа не оторвалась от истоков нацистского учения». А как же тогда быть со «священными камнями Европы» Достоевского? Полагаю, что не «Бог умер!», как философ Ницше уверенно полагал, а антураж его померк для тех, кто Бога забыл. Но имманентно святая сущность Европы живёт себе в нуждающихся в ней. Мир, по философу Артуру Шопенгауэру - «воля и представление». Спрашивается: «фашистский» ли донос на себя осуществил большой поэт Владимир Маяковский: «Я люблю смотреть, как умирают дети»? Так что «не сравнивай: живущий не сравним».

24. Сложна и не изучена феноменология сего «кентаврова» сращения светлого и тёмного миров в человеке. Полагаю: есть ситуационная трансформация, вынужденная, поверхностная, с внешним уподоблением – временная аномалия и полное перерождение в исчадие ада - отпетого негодяя. По философу Виталию Даренскому, сложившаяся укропостсоветская национальная «недоидентифицированность» - синром несостоявшейся государственности (a failed nation). Мотив - «мстительность» («ressentiment»).

25. Поясню. В отличие от «недоидентичности», идентичность - устойчивая платформа ценностей, верований и убеждений, определяющих как повседневное поведение людей, так и их долговременных жизненных программ. Надломив общее с Россией русско-советское корневище, тем самым выйдя из реальной идентичности общесоветского приобщения в том числе и украинского этноса к большой истории СССР-России, *украинцы «пытаются выстроить псевдоидентичность из обломков племенного сознания, с одной стороны, и из поверхностно-эпигонского усвоения элементов «европеизма»*, с другой. В ре-

зультате утопичности этих попыток возникает *симуляция идентичности **при её реальном отсутствии***. И именно в этом кроется *причина агрессивной русофобии современного «украинства»*. За ним стоит *«бессознательное чувство вины за предательство своей реальной, а не выдуманной общности, созданной русской историей» «триипостасного» народа (русских, украинцев и белорусов). Период самостийности выявил тотальную деградацию «неоукров» (как, впрочем, и «новых русских»)* То есть, «нацификация» - расплата за обрывание пуповины общей социо-генетической кровной судьбы с Россией. Назову эту этико-мировоззренческую аномалию *синдромом утраченной идентификации и попыткой обретения взамен чужого как своего.* По поведенчески аномальному сходству с «германскими нацистами», при характеристике украинских радикалов в России используется термин «нацизм» (прецедентная формулировка приговора Международного военного трибунала в Нюрнберге от 1 октября 1946).

26. Концепт Даренского интересен в плане выяснения возможного воздействия на сознание и поведение фактора *«недо*идентифицированности» или

*«псевдо*идентифицированности» (ложное альтер-эго). Симуляция идентичности носит компенсационный характер (до целостной полноты её). Вместе с тем, вряд ли правомерно исходить в оценке сего феномена из «***реального отсутствия***» идентификации как факта. Полагаю: она остаётся, но в искажённом виде, как коллаж наложения коррективов новизны. При чем а-ля «нацистская» привнесённая атрибутика оценивается как критериальная доминанта нравственного облика украинского сопротивленца. По философу Джону Локку идентичность личности состоит в тождестве непрерывного сознания, а не какой-либо субстанции (вне зависимости от того материальная она или нет) - «одну и ту же личность образует *не одна и та же субстанция*», а одно и тоже *непрерывное сознание.* Согласно психосоциальному пониманию самоидентичности, идентичность представляет собой ***поддержание баланса между внутренней целостностью и ценностями общества.*** Мыслитель Михаил Бахтин считал, что *настоящее «Я» всегда обнаруживается в точках несовпадения человека с самим собой, в его идентификациях с «Другим».* Следовательно, НОВАЯ САМОИДЕНТИФИКАЦИЯ могла уже взрости до

равнозначного «другого Я», полноценного в смысле полноты формирования Alter ego. То есть, кажущийся мутантом – вполне целен в зеркальном отражени-и своего «Я». Первоархетипы смутно-иного дремлят в нём, но обновлённая до неузнавания переиначенная сущность вполне состоялась: синдром «банде-ризма» генно дремал априори, в ВОВ прошёл закалку огнём и кровью, пережил годы гоне-ний (тюрьмы, ссылки), ожил в хрущёвскую «оттепель», пустил корни и предстал в пост-Бело-вежской России и реализовался в схлёстке ныне с «оккупантами самостийной» (его второго «Я»).

27. Да и сама «политика идентичности» даёт возможность влиять человеку на информацию о себе, которая поступает в социум. И наконец, модно в определённых стратах иметь «крутые» наколки-тату. Посему смежные этносы одного народа образуют единство противоположностей: «брат мой – враг мой». «Украина всегда была биполярной (оранжево - голубая), в ней **расправы наглядны и публичны**. Россия - более тонкая и глобальная си-стема, связанная как с условным глобальным Киссинджером, так и с условным глобальным Со-росом теснее Украины» - пишет М.Шевченко. «Ни прозвание, ни вероисповедание, ни самая кровь предков не делают чело-

века принадлежностью той или другой народности. Дух, душа человека – вот где надо искать принадлежности его к тому или другому народу. Чем же можно определить принадлежность ду-ха? Конечно, проявлением духа – мыслью. Кто на каком языке думает, тот к тому народу и при-надлежит. Я думаю по-русски», – писал Владимир Даль (родился на Луганском горном заводе, в нынешнем Донбассе). ***Так что даю новый в социальной науке аспект:*** *специальная военная операция как коррекция двух аномалий: «бандеровской» самоидентификации Украины и государства Российского как госкорпорации. Избавление от «новой русскости», как и «укронацизма» ради возвращения к своим первородным этно-обликам - будущее единого трёхипостасного народа.*

28. Прикрываться заложниками – это не только тактика террористов, но и стратегия давних баталий. Да и сам по себе *«живой щит» ведь может быть и на добровольных началах* актом самопожертвования для поддержания военных усилий. По зову патриотического сознания – помогать Отечеству в опасности чем можно, «не щадя живота своего». В мире перевёрнутых морально-этических ценностей, законничества и формализованного

фрагментарного сознания – нетрадиционное сразу объярлычивается: «терроризм»-«ИГИЛ»-«нацизм». Так что без суда и следствия, нельзя гуртом зачислять эту элиту украинского сопротивления в «укронацистов», а затем столь же немотивированно – разом обелять всех: мол *только иноземное* начальство воинов – подонки и мразь.

29. Нужна и в РФ, в Украине «дезактивация» и «дезинфекция» больного социума. Война –индикатор, «оселок» гражданской полноценности и стимулятор здорового духа нации. Могильная плита уравнивает *всех* комбатантов –защитников своей Родины. А ультрапатриоты –- хулимые в РФ как «националисты», уподобляемые «нацистам», -- нередко настоящие рьяные патриоты, в большинстве своём бесстрашные воины – «квасные патриоты». Не говорю о садистах-мучителях, патологических извращенцах. Не исключено, что «горячим парням» с обеих сторон – националистам «украм» и по-аввакумовски неистовым «русопятам» (под кликухой «краснокоричневые») - предстоят ратные дела по расчистке общего дома и воссозданию единого народа. Ведь национальное унижение дошло до того, что премьер

Польши Матеуш Моравецкий смеет открыто заявить о необходимости *уничтожить «русский мир»*. Победой в спецоперации многократно усилится Русский мир, а поражение в ней - загонит в подполье общеславянское имперское возрождение. Это экзистенциальная операция – быть или не жить. «А зачем нам такой мир, если там не будет России?» - Путин сформулировал ключевую фразу русского самосознания.

30. Без подпитки почвой не будет окрыляющего победного духа. Однако пуповина, связующая воинство и почву, в России надорвана вместе с заколоченными окнами деревень. Есть угроза превратиться в «Иванов, не помнящих родства». Или «манкуртов» (словечко из романа Чингиза Айтматова «Буранный полустанок») - полностью утративших связь со своими историческими, национальными корнями, забыв о своём родстве. Менталитет формирует чувство Родины, является фундаментом национального самосознания.

31. Владимир Путин – чёрнопоясник по дзюдо, поистине супермен традиции в аномальном мире перевёрнутых с ног на голову ценностей. Если у него и «склерозит» принятие

решений, то он ничуть не синельно-диментен, не при «смерти мозга», не спятил ни в коммунизм, ни в нацизм. Клинически не аномален. Поведенчески не «садомазен» (в ответных санкциях). Как дзюдоист, он ждёт ошибки «партнёра» - чтобы бросить на ковёр. Запад, вынуждая Россию если не трансформироваться, то «самоликвидироваться», самурайски харакирит себя: подталкивает Россию к стратегической спарке с Китаем, даже рискует превратить США в Разъединённые Штаты Америки... Но «Стратегия Анаконды» не дремлет: она в разгаре третьей удушающей РФ фазы – «ликвидации Российской Федерации в её нынешних границах, с полной утратой экономичечского, политического и военного суверенитета».

32. При всём при том, козыри биты или репутационно подмочены: благоприятный момент *допереворота* (в Беловежье был, полагаю, *недоворот*) в России, пожалуй, упущен. Оп-позиция обезглавлена. Михаил Ходорковский - за кордоном. Алексей Навальный - в тюрьме. Анатолий Чубайс, с прожектом реванш-либеральной диктатуры, – смылся с политико-идеологи-ческого фона РФ. Дипломатия сдохла - оружие апокалипсиса торпеда «Посейдон» и МБР РС-28 «Сарматы» наготове. А сия

десятитонная ракетища может доставить смертоносный груз даже через Южный полюс (он ведь почти не контролируется) и способна, как авторитетно заявил глава «Роскосмоса» Дмитрий Рогозин, уничтожить половину берега континента. А показ Западу воронки от ракеты «Сармат», как он полагает, - лучшая рекомендация быть вежливее с Россией.

33. Таким образом, экспозиция расстановки сил в необъявленной войне между Западом и Россией такая: Вашингтон (ЕС и Зеленский) vs. Россия (в одиночку, с «пятой колонной» в тылу у себя, при некоторой «асимметрии» в принятии решений и невнятице их толкования). На Украине по сути Гражданская война скованных одной цепью кровно родственных славян: традиционалистов (как советские шахтёры под знаменем общей победы в ВОВ) и неофитов (широкомасштабно переформатировавшихся в пробандеровски самоидентифицированное народонаселение). Столь неистово остервенело противоборствовали Красные «террористы» и Белогвардейские застеночники. Но отделим «котлеты от мух»: Однако, кроме извергов по натуре (подлецов по рождению – где-то у раннего Достоевского под-

метил), может быть и просто ультра-националистический элемент, который образовался вследствие сложившихся негативных обстоятельств. Например: проявление национального достоинства, долго сдерживаемое властями и квалифицируемое ими как незаконное, способно «экстримально» вырваться на волю, как джин из бутылки, и мстить миру за своё мучительное прошлое. Претерпевшие моральное надругательство, нравственное истязание интеллигенты нередко пребывают в личине горько «смеющихся сквозь слёзы» - синдром мстящего «раненного» сознания. (Я копался в этих штудиях в давнюю бытность диссертантом-соискателем, с темой «Проблема амбиции у Достоевского как фактор русского сознания в принятии решений»).

34. В «лихие 90-е» подлинные патриоты России оказались лишними в своём Отечестве (и ваш покорный слуга в том числе): ненужными установившемуся в РФ антинациональному режиму «криминал-революционеров». «Абортированные» Россией (изгнанные за рубеж), они оказались востребованными в частности антирусским олигархатом «незалежной». И по сей день «нацики» - оплот отпетого олигархического

киевского режима – заградотрядники «ни шагу назад!».

35. Дожать Россию – момент упущен. Наблюдающаяся в последнее время активизация российских военных усилий на украинском направлении побудила опытнейшего дипломата экс-госсекретаря США Генри Киссинджера неотложно рекомендовать скорейший мир: «Запад должен отказаться от попыток добиться военного поражения России в Украине, а последней стоит пойти на территориальные уступки». Видимо, чтобы превентивно застопорить закрепление русского нарастающего успеха. Чтоб оставить при РФ лишь ограниченные нынешние территории, отвоёванные у Украины (как войска Жукова и Паттона спешили навстречу друг другу, и в результате - встретились на Эльбе). И только. Что позволило бы эту «пиррову победу» обернуть через небольшой промежуток временной паузы в тотальное поражение России. Жертвуя частью – спасаешь целое (логика ленинского «Брестского похабного мира»).

36. Но политической бюрократии не нужна *ничья* скорая победа. Поскольку это неизбежно повлечёт за собой смену

обанкротившихся «элит». Поэтому саботажем выхода на подписание акта о капитуляции Украины, с обеих сторон будет поддерживаться сценарий войны «до победного конца». Политбюрократии «нужна не конкретная военная победа, а её вечное предчувствие». Посему «глобальная гибридная» затянется «до последнего украинца» или «доллара в казне США». Даже для опытнейшего генерала Владимира Шаманова вопрос о будущем Украины труднопрогнозируем. Можно лишь предполагать, что демилитаризация страны может занять от пяти до десяти лет, а сформировать там власть, «не запачканную неонацистами», будет сложно. При этом он считает, что, если этих двух явлений не произойдёт, Украина не подойдёт России в качестве нового субъекта федерации.

37. Но рано делить шкуру неубитого медведя и нелепо – если это кровный родич. Но ради одного *трёхипостасного* русского народа тогда отрадно. Патриот Дмитрий Рогозин только что предложил «наладить жизнь» на космических заводах Украины. Когда мир был двуполярным, американо-российским, глобальная безопасность держалась в понятных всем пределах: ни войны – ни мира. Не к этому ли придём? Иначе вызов Китая унесёт в небытие

цивилизационный конфликт США и РФ.

38. Высшая номенклатура и России, и Украины пытается всеми силами сорвать эту идущую «концептуально очистительную» войну (ВСО). Практически остановка продвижения войск — это следствие ДОГОВОРНЯКА с «бандеровцами». Министр обороны РФ С.Шойгу объясняет замедление наступления российских войск на Украине лишь стратегическим манёвром, гарантирующем минимизацию потерь среди гражданского населения Украины. Но такая тактика противоречит азам военного искусства. Ибо непреложна истина: высокие темпы наступления ведут к существенному снижению потерь наступающих и к высоким потерям обороняющихся. Неужто эту аксиому «науки побеждать» забыл Генштаб, назубок знающий монографию теоретика В.Савкина "Темпы наступления"? Или же военные осуществляют в точности в пределах дозволенного предписанную им стратегию тайного сговора правящих верхов обеих стран? Ведь в этом пособии для генералитета ясно сказано: «При высоких темпах продвижения и эффективном подавлении противника огнём наступающий несёт меньшие потери в живой силе и технике. С повышением темпов наступления происхо-

дит дальнейшее уменьшение потерь». За полгода боев российскую армию сознательно сковывали - мешали выйти на границы Приднестровья, водрузив флаг над Одессой. Такие правящие режимы, вольно или неволь множащие потери с обеих сторон (одни - деляческой прагматикой, другие - безмозглым неистовством «до последнего украинца», причём без экипировки), надлежат смене на национально «вменяемых» (власть умеренных националистов). Война очистит национальные горизонты и может перерасти в обеих странах в войны гражданские - «за нашу и вашу свободу». Ну а совместно ли биться с наднациональным олигархатом - решат сами стравленные народы. В этом и состоит экзистенциальный смысл военно-ментальной сшибки в восточно-славянском этносе.

Д-р Евгений Александрович Вертлиб/Dr. Eugene Alexander Vertlieb президент Международного института стратегических оценок и управления конфликтами (МИСОУК-Франция); ответственный редактор отдела прогнозирования политики Запада «Славянской Европы» (Мюнхен); экзекьютив член Инициативы «Лиссабон-Владивосток» (Париж)

Border Security: Vulnerability and Emerging Global Security Threats

Robert Girod

ABSTRACT

National sovereignty and global alliances face new challenges due to evolving and emerging security threats, ranging from pandemic "vaccine nationalism" to viral cyber-attacks. While pandemics and cyber-attacks are global threats that cross international borders, border sovereignty is still and increasingly a threat to national security and homeland security that takes the form of espionage, terrorism, narcotics and dangerous drugs trafficking, human trafficking, weapons and other contraband smuggling, counterfeiting and intellectual property theft, economic crimes, and a plethora of other threats. Not only are criminals, terrorists, and agent provocateurs a threat to borders, but nation-state opponents are threats to border sovereignty and the subject of international disputes that verge on the edge of war.

Keywords: Border Security; vulnerabilities; Ukraine War; National Sovereignty

Seguridad fronteriza: Vulnerabilidad y amenazas emergentes a la seguridad global

RESUMEN

La soberanía nacional y las alianzas globales enfrentan nuevos desafíos debido a las amenazas a la seguridad emergentes y en evolución, que van desde el "nacionalismo de las vacunas" pandémico hasta los ciberataques virales. Si bien las pandemias y los ataques cibernéticos son amenazas globales que cruzan fronteras internacionales, la soberanía fronteriza sigue siendo, y cada vez más, una amenaza a la seguridad nacional y a la seguridad nacional que toma la forma de espionaje, terrorismo, tráfico de narcóticos y drogas peligrosas, trata de personas, armas y otros tipos de contrabando. contrabando, falsificación y robo de propiedad intelectual, delitos económicos y una gran cantidad de otras amenazas. Los criminales, los terroristas y los agentes provocadores no sólo son una amenaza para las fronteras, sino que los oponentes de los Estados-nación son amenazas a la soberanía fronteriza y son objeto de disputas internacionales que

doi: 10.18278/gsis.8.1.10

rozan el borde de la guerra.

Palabras clave: Seguridad de frontera; vulnerabilidades; Guerra de Ucrania; Soberanía nacional

边境安全：脆弱性与新兴的全球安全威胁

摘要

由于不断变化和新兴的安全威胁（从大流行"疫苗民族主义"到病毒式网络攻击），国家主权和全球联盟面临着新的挑战。尽管大流行和网络攻击是跨越国界的全球性威胁，但边境主权仍然对国家安全和国土安全构成越来越大的威胁，其形式包括间谍活动、恐怖主义、麻醉品和危险毒品贩运、人口贩运、武器和其他违禁品走私、伪造和知识产权盗窃、经济犯罪、以及大量其他威胁。对边境构成威胁的不仅有犯罪分子、恐怖分子和诱导者，而且民族国家的反对者也对边境主权造成威胁，他们也是濒临战争边缘的国际争端的主题。

关键词：边境安全，脆弱性，乌克兰战争，国家主权

Introduction

The United States is not the only country with border security issues. The United States and Canada have maintained friendly relations for more than a century, but border disputes between these two allied countries still exist. The United States and Canada have at least five current and ongoing border disputes, each involving maritime claims. Many territories around the globe are claimed by more than one country, often leading to tensions and conflict between sovereign countries, sometimes resulting in war. Globally, there are (at this time) more than 150 territorial disputes between nation-states. China has the largest number of neighbors (fourteen) sharing its 22,000 km land borders: North Korea, Russia, Mongolia, Kazakhstan, Kyrgyzstan, Tajikistan, Afghanistan, Pakistan, India, Nepal, Bhutan, Myanmar, Laos and Vietnam and is engaged in territorial disputes with all fourteen of these countries.

The worldwide press and media have been reporting on many border issues that have led to global insecurity and conflicts. Illegal immigration, COVID and similar pandemics, arms and drug trafficking, human trafficking and other criminal activity, territorial disputes, and cyber-attacks are only a few of the issues affecting border security and the resulting vulnerabilities and emerging global threats to national and multi-national security interests.

Border security is an essential element of both national security and homeland security, in that "border security" secures a nation-state from threats to national defense and from foreign intelligence activities and homelands from public safety and law enforcement threats. The former includes potential espionage, sabotage, and terrorism, while the latter includes acts of human trafficking, narcotics and dangerous drug offenses, the smuggling of weapons and other contraband, and counterfeit or prohibited items. Borders also include boundaries created in cyberspace by malware, viruses, and hacking. Cyber viruses are not the only viruses of concern to border security; pandemic viruses and biological warfare are also homeland and national security issues of concern.

Global Alliances Face New Challenges

Border disputes is only one of the threats faced by nation-states alliances. The Five Eyes (FVEY) alliance is an intelligence alliance comprising Australia, Canada, New Zealand, the United Kingdom (UK), and the United States (US). These countries are parties to the multilateral **UKUSA Agreement**, a treaty for joint cooperation in signals intelligence and, more informally, the group of intelligence agencies represented by these countries. Other alliances, such as NATO and SEATO, face numerous mutual defense and security challenges. But such intelligence, defense, and security alliances are not the only global interests.

The world-wide pandemic has brought new challenges to a variety of organizations, affecting not only world health issues, but border security, sovereignty legal issues, and threats to national security. The World Health Organization (WHO), for example, is a specialized agency of the United Nations (U.N.) responsible for international public health. The WHO is the U.N. agency that "connects nations, partners and people to promote health, keep the world safe, and serve the vulnerable." (World Health Organization, 2022)

WHO's Global Health Estimates (GHE) provide the latest available data on death and disability globally, providing insight to support informed decision-making on health policy and resource allocation. Public health surveillance is the continuous, systematic collection, analysis and interpretation of health-related data. Humanitarian emergencies increase the risk of transmission of infectious diseases and an effective disease surveillance system is essential to detecting disease outbreaks quickly before they spread, cost lives and become out of control. (World Health Organization, 2022)

National Sovereignty and the Nature of Border Security and Barriers

National Sovereignty is, in part, defined by its borders, its citizens, and its laws. Professors James R. Phelps, Jeffrey Dailey, and Monica Koenigsberg concisely observe that, "How effectively border integrity and security are provided directly affects all citizens, legal residents, and those who want to profit from violating internationally accepted rules on national sovereignty." (Phelps, pp. 3-4) Borders, boundaries, and barriers are more than lines on a map. They reinforce these political demarcations and require an understanding of a wide variety issues that affect the human condition around the world. Borders are the edge or periphery used to delineate national and political boundaries. Boundaries are anything that indicates or fixes political, economic. Legal, physical, or even mental limits. Barriers are any material—natural or manufactured—that blocks, prevents or hinders passage across borders or boundaries.

The most common barriers are walls. The most iconic and probably oldest barrier wall is the Great Wall of China—a series of fortifications and a wall made of stone, brick, tamped earth, wood, and other materials—which runs east and west through the entire northern part of China. The Great Wall was built between 220 and 206 B.C.E. by Qin Shi Huang, the first Emperor, to prevent intrusions and military incur-

sions by nomads from Mongolia and Siberia. Now a tourist attraction that can be seen from space, the Great Wall was a means of border control, checkpoints, and customs collection centers. (Phelps, p. 26)

The Maginot Line was a defensive line built between France and Germany (3-6 miles back from the German frontier) consisting of pillboxes or fortified houses, minefields and anti-tank traps. (Phelps, p. 53) Meanwhile, the Germans also built the Siegfried Line to defend against attacks from its eastern borders. (Phelps, p. 55) Following World War II, "The Iron Curtain" was a physical fence built to keep people in and separating Eastern and Western Europe. "The Berlin Wall" represented this "Iron Curtain" along the international borders between Warsaw Pact and North Atlantic Treaty Organization (NATO) countries. (Phelps, pp. 57-58) (Phelps, pp. 3-4) Again, this series of walls and fences differed from others, in that it was designed to keep people in—to prevent them from fleeing Soviet-dominated tyranny.

Land borders are not the only concern in the protection of national borders. Maritime borders are just as vulnerable, if not more so, because of their permeable nature. The *coast* is determined by a "smoothed" line running across inlets, bays, channels, harbors, etc. The "shore" is an actual point where land meets ocean saltwater. The difference is merely a matter of measurement—the United States has 12,380.2 miles of "coastline" but 95,000 miles of "shoreline." (Phelps, p. 151)

The major concern with ocean borders is the safety and security of maritime commerce, which may be threatened by vulnerability of the supply chain, human error, corruption, and compromise. (Phelps, p. 190)

Transnational Threats to Cyber Security

Cyber threats have evolved from lone hacker crimes to transnational terrorism and espionage threats. Rising ransomware attacks against critical infrastructure have shown that threats have crossed digital and physical borders. The decentralization of the workplace makes endpoint security even more critical than ever. New tactics used by malicious actors require focus on different tools and solutions. The U.S. Department of Homeland Security (DHS) and the U.S. Department of Commerce's National Institute of Standards and Technology (NIST) have released standards to help organizations protect their data and systems from risks associated with the advancement of quantum computing technology. While quantum computing promises unprecedented speed and power in computing, it also poses new risks. As this technology advances, it is expected to break encryption methods that are widely used to protect customer data, complete business transactions, and secure communications. New guidelines will help organizations prepare for the transition to post-quantum cryptography by identifying, prioritizing, and protecting potentially vulnerable data, algorithms, protocols, and systems. (Department of Homeland Security, 2022)

Cyberspace and its underlying infrastructure are vulnerable to a wide range of risks stemming from both cyber and physical threats. Sophisticated cyber actors and nation-states exploit vulnerabilities to steal information and money and are developing capabilities to disrupt, destroy, or threaten the delivery of essential services. Cyberspace is particularly difficult to secure due to a number of factors: 1) the ability of malicious actors to operate from anywhere in the world, 2) the linkages between cyberspace and physical systems, and 3) the difficulty of reducing vulnerabilities and consequences in complex cyber networks. Of growing concern is the cyber threat to critical infrastructure and cyber intrusions. As information technology becomes increasingly integrated with physical infrastructure operations, there is increased risk of disruption of services. (Cyber Security & Infrastructure Security Agency, 2022)

Cyber threats are simultaneously a national security and homeland security threat and a counterintelligence problem. State- and non-state actors use digital technologies to achieve economic and military advantage, instigate instability, increase control over content in cyberspace and achieve other strategic goals—often faster than our ability to understand the security implications and neutralize the threat. (Office of the Director of National Intelligence, 2022) Some examples of cyber-attacks include:

- In Estonia, in 2007, during a period of political tensions between the

Russian Federation and Estonia, there were a series of denial-of-service (DOS) cyberattacks against many Estonian websites, including those run by the Estonian Parliament, government ministries, banks, newspapers and television stations. Though Russia was blamed for these attacks based on circumstantial evidence, the Russian Government never admitted its involvement. An ethnic Russian living in Tallinn, who was upset by Estonia's actions and who had been acting alone, was convicted in an Estonian court for his part in these attacks.

- In Dharamsala, India, in 2009, security researchers discovered a sophisticated surveillance system in the Dalai Lama's computer network, called *GhostNet*. The same network had infiltrated political, economic and media targets in 103 countries. China was the suspected origin of this GhostNet, based only upon circumstantial evidence. It was also unknown whether this network was run by a government organization or by Chinese nationals for either profit or nationalist reasons.

- In Iran, in 2010, the *Stuxnet* computer worm severely damaged, and possibly destroyed, centrifuge machines in the Natanz uranium enrichment facility, in an effort to set back the Iranian nuclear program. Analysis of the worm indicated that it was a well-designed and well-executed cyber-weapon, which required an engineering effort that

implied a nation-state sponsor. Investigative reporting suggested that the United States and Israel were the designers and deployers of the worm, although neither country has officially taken credit for it.

(UN Chronicles, 2022)

IP Theft and Propaganda

CNN reports that the CIA is revamping how it trains and manages its operatives as part of a "broader effort to transition" away from 20 years of counterterrorism wars and focus more closely on adversaries like China and Russia. After two decades of paramilitary operations against Islamist terror groups, some former intelligence officers say the CIA needs to get back to "traditional, quiet tradecraft" to collect intelligence on complex nation-states, such as China, which senior officials say presents the agency with its biggest challenge. (CNN Politics, 2021)

John Lenczowski, PhD, former NSC Staff Expert, and currently Chancellor of the Institute of World Politics, says that while the psychological disarmament of the decision-making elites in the U.S. and Western allies has led to the global theft of intellectual property, the Chinese use the media, academic institutions, think tanks, business community, politicians, and the entertainment industry as a tool for their massive propaganda operations. Many academicians, universities, research enterprises have taken large grants from China, or done joint projects with Chinese front organizations. As a result, an enormous amount of the reports which emanate from these enterprises are skewed to follow Chinese talking points on major issues. (AFIO, 2022)

The Washington Post reported on January 17, 2022, that the Chinese Olympic organizing committee have warned that foreign athletes may face punishment for speech that violates Chinese law at the 2022 Winter Games. (Eva Dou, Washington Post, 2022)

COVID Pandemic as a Transnational Security Threat

COVID-19 is a nontraditional threat which proliferates with or without intention, bargaining or goals. This nontraditional threat has affected U.S. national security in both direct and indirect ways. This is both a domestic issue and an international issue. Whether pandemic viruses are weaponized WMDs or not, such public health crises are clearly threats to both national security and homeland security. Different countries have their own entry and exit requirements and the U.S. Center for Disease Control (CDC) recommends checking each country's requirements before planning international travel.

Migration at the Polish-Belarusian Border

National Geographic reported on January 31, 2022, that the once peaceful countryside border between Poland and Belarus of forests, rolling hills, river valleys, and wetlands, has become a militarized zone. Concerns about an influx of primarily Middle Eastern migrants from Belarus, have prompted the Polish government to construct a massive wall across its eastern border. (Douglas Main, National Geographic, 2022) The Belarusian-Polish border is the state border between the Republic of Poland and the Republic of Belarus. It is between 248 miles (398.6 km) and 260 miles (418 km) long, beginning at the triple junction of the borders with Lithuania in the north and stretching to the triple junction borders with Ukraine to the south.

Aljazeera also reported on the Polish border wall, saying, "Poland branded the crisis a 'hybrid' attack from Belarus and its main ally Russia, referring to a type of warfare using non-military tactics. Belarus has denied this and has, in turn, accused Poland of "inhumane treatment of the refugees." Belarus has since repatriated thousands of people to Iraq, the primary origin of the refugees. At the height of the crisis, Poland sent thousands of troops and police officers to the border to reinforce border guard patrols. (Aljazeera. 2022)

On 02 September 2021, Poland's government announced a state of emergency along its border with Belarus as a result of the ongoing migration and security crisis and established an exclusion zone along the frontier that has barred journalists, relief groups, and nonresidents from accessing the area. But new regulations were implemented on 01 December 2021 and the Polish Border Guard has begun granting journalists limited access to the border zone as part of supervised tours. *Although soldiers keep the peace, fences damaged by Belarusian guards to allow groups of migrants to cross the border, creating security challenges that Belarus has presented to Poland's sovereignty.*

National Sovereignty, Global Alliances, and the Russian-Ukraine War

Russia has history of resisting Ukraine's move towards European institutions and the **North Atlantic Treaty Organization** (NATO), a defensive alliance of 30 countries[1] (an intergovernmental military alliance between 27 European countries, 2 North American countries, and 1 Eurasian country). Russia's demand is for the West to guarantee Ukraine will not join NATO. Ukraine shares borders with both the **European Union** (EU) (a political and economic union of 27 member states that are located primarily in Europe) and Russia. But as a former Soviet republic Ukraine has deep social and cultural ties with Russia and Russian is widely spoken there.

The threat is being taken seriously because Russia has invaded Ukraine before. When Ukrainians deposed their pro-Russian president in early 2014, Russia annexed Ukraine's southern Crimean peninsula and backed separatists who captured large areas of

eastern Ukraine. The rebels have fought the Ukrainian military ever since in a conflict that has claimed more than 14,000 lives. The BBC reports conflicting opinions and rhetoric on the situations: "Russia says it has no plans to attack Ukraine: and armed forces chief Valery Gerasimov even denounced reports of an impending invasion as a lie." And while NATO's Secretary General warns that the risk of conflict is real, "U.S. officials" have stressed "they do not believe Russia has decided on an invasion, and it is not imminent." Meanwhile, Ukraine's president has appealed to the West not to spread "panic." And still, Russian President Vladimir Putin has threatened "appropriate retaliatory military-technical measures" if what he calls "the West's aggressive approach" continues. (Paul Kirby, BBC News, 2022)

Unidentified U.S. sources report that Russia has offered no explanation for the troops posted close to Ukraine and that thousands of Russian troops have headed to Belarus, close to Ukraine's northern border, for "exercises." But Russia's deputy foreign minister compared the situation to the 1962 Cuban missile crisis, when the U.S. and Soviet Union came close to nuclear conflict. Western intelligence opine a Russian incursion or invasion could happen sometime in early 2022. Why is Russia insistent that Ukraine not join NATO? What is its security in-

---

1   Albania, Belgium, Bulgaria, Canada, Croatia, Czech Republic, Denmark, Estonia, France, Germany, Greece, Hungary, Iceland, Italy, Latvia, Lithuania, Luxembourg, Montenegro, Netherlands, North Macedonia, Norway, Poland, Portugal, Romania, Slovakia, Slovenia, Spain, Turkey, United Kingdom, United States.

terest? Russia has spoken of a "moment of truth" in its relationship with NATO. Deputy Foreign Minister Sergei Ryabkov said, "For us it's absolutely mandatory to ensure Ukraine never, ever becomes a member of NATO." And President Putin explained that if Ukraine joined NATO, the alliance might try to recapture Crimea. Moscow accuses NATO countries of "pumping" Ukraine with weapons and the U.S. of stoking tensions to contain Russia's development. (Paul Kirby, BBC News, 2022)

In short, Russia would like NATO's borders to return to pre-1997 boundaries. It demands no more eastward expansion and an end to NATO military activity in Eastern Europe. That would mean combat units being pulled out of Poland and the Baltic republics of Estonia, Latvia and Lithuania, and no missiles deployed in countries such as Poland and Romania. Russia has also proposed a treaty with the U.S. barring nuclear weapons from being deployed beyond their national territories. This would have immense strategic implications for U.S. deterrent capabilities. Meanwhile, the Pentagon has accused Russia of preparing a so-called false-flag operation, with operatives ready to carry out acts of sabotage against Russian-backed rebels, to provide a pretext for invasion. Russia, of course, has denied this.

Armed Forces of the Russian Federation:[2]

- Ground Troops (Sukhoputnyye Voyskia, SV)

- Navy (Voyenno-Morskoy Flot, VMF)

- Aerospace Forces (Vozdushno-Kosmicheskiye Sily, VKS)

- Airborne Troops (Vozdushno-Desantnyye Voyska, VDV)

- Missile Troops of Strategic Purpose (Raketnyye Voyska Strategicheskogo Naznacheniya, RVSN) referred to commonly as Strategic Rocket Forces (independent "combat arms," not subordinate to any of the three branches

- Federal National Guard Troops Service of the Russian Federation (National Guard (FSVNG), Russian Guard, or Rosgvardiya): created in 2016 as an independent agency for internal/regime security, combating terrorism and narcotics trafficking, protecting important state facilities and government personnel, and supporting border security; forces under the National Guard include the Special Purpose Mobile Units (OMON), Special Rapid Response Detachment (SOBR), and Interior Troops (VV); these troops were originally under the command of the Interior Ministry (MVD)

(CIA World Factbook, 2022)

---

2   CIA World Factbook (Russia)

Federal Security Services Border Troops (includes land and maritime forces) (2021)[3]

The Air Force and Aerospace Defense Forces were merged into the VKS in 2015; VKS responsibilities also include launching military and dual-use satellites, maintaining military satellites, and monitoring and defending against space threats.[4] (CIA World Factbook, 2022)

Information varies; there are approximately 850,000 total active duty troops:[5]

- 375,000 Ground Troops, including about 40,000 Airborne Troops;

- 150,000 Navy;

- 160,000 Aerospace Forces;

- 70,000 Strategic Rocket Forces;

- 90,000 other uniformed personnel (approximately 20,000 special operations forces, plus command and control, cyber, support, logistics, security, etc.); and

- 200,000-250,000 Federal National Guard Troops (2021).

(CIA World Factbook, 2022)

The Russian Federation's military and paramilitary services are equipped with domestically-produced weapons systems, although since 2010 Russia has imported limited amounts of military hardware from several countries, including Czechia (The Czech Republic), France, Israel, Italy, Turkey, and Ukraine. Yet the Russian defense industry is capable of designing, developing, and producing a full range of advanced air, land, missile, and naval systems; Russia is the world's second largest exporter of military hardware (2021).[6] (CIA World Factbook, 2022)

Human Trafficking: Russia is a source, transit, and destination country for men, women, and children who are subjected to forced labor and sex trafficking, although labor trafficking is the predominant problem; people from Russia and other countries in Europe, Central Asia, Southeast Asia and Asia, including Vietnam and North Korea, are subjected to conditions of forced labor in Russia's construction, manufacturing, agriculture, repair shop, and domestic services industries, as well as forced begging and narcotics cultivation; North Koreans contracted under bilateral government arrangements to work in the timber industry in the Russian Far East reportedly are subjected to forced labor; Russian women and children were reported to be victims of sex trafficking in Russia, Northeast Asia, Europe, Central Asia, and the Middle East, while women from European, African, and Central Asian countries were reportedly forced into prostitution in

---

3   CIA World Factbook (Russia)

4   CIA World Factbook (Russia)

5   CIA World Factbook (Russia)

6   CIA World Factbook (Russia)

Russia.[7] (CIA World Factbook, 2022)

Armed Forces of Ukraine (Zbroyni Syly Ukrayiny, ZSU):[8]

- Ground Forces (Sukhoputni Viys'ka),

- Naval Forces (Viys'kovo-Mors'ki Syly, VMS),

- Air Forces (Povitryani Syly, PS),

- Air Assault Forces (Desantno-shturmovi Viyska, DShV),

- Ukrainian Special Operations Forces (UASOF),

- Territorial Defense Forces (Reserves).

Ministry of Internal Affairs: National Guard of Ukraine, State Border Guard Service of Ukraine (includes Maritime Border Guard) (2021)[9] (CIA World Factbook, 2022)

Information varies; there are approximately 200,000 active troops:[10]

- 150,000 Army, including Airborne/Air Assault Forces;

- 12,000 Navy;

- 40,000 Air Force; and

- approximately 50,000 National

Guard (2021).[11]

(CIA World Factbook, 2022)

A Recap of Chinese-Taiwan History and Conflict

Taiwan, officially the Republic of China (ROC), in East Asia, shares maritime borders (a conceptual division of the Earth's water surface areas using physiographic or geopolitical criteria) with the People's Republic of China (PRC) to the northwest, Japan to the northeast, and the Philippines to the south. The main island of Taiwan, formerly known as *Formosa*, has an area of 35,808 square kilometers (13,826 square miles), with mountain ranges dominating the eastern two-thirds and plains in the western third. The capital is Taipei, which, along with New Taipei City and Keelung, Makes up the largest metropolitan area of Taiwan.

It is believed that Taiwan was settled by farmers, most likely from what is now southeast China, around 6,000 years ago. Han Chinese fishermen began settling in the Penghu islands in the 13th century. The Dutch East India Company attempted to establish a trading outpost on the Penghu Islands (Pescadores) in 1622, but was driven off by Ming forces. In 1626, the Spanish Empire landed on and occupied northern

---

7   CIA World Factbook (Russia)

8   CIA World Factbook (Ukraine)

9   CIA World Factbook (Ukraine)

10   CIA World Factbook (Ukraine)

11   CIA World Factbook (Ukraine)

Taiwan as a trading base, first at Keelung and in 1628 building Fort San Domingo. This colony lasted 16 years until 1642, when the last Spanish fortress fell to Dutch forces. After being ousted from Taiwan, the Dutch allied with the new Qing dynasty in China against the Zheng regime in Taiwan. In 1683, the Qing dynasty formally annexed Taiwan. Following Qing's defeat in the First Sino-Japanese War (1894–1895), a conflict between the Qing Dynasty of China and the Empire of Japan,, Taiwan, its associated islands, and the Penghu archipelago were ceded to the Empire of Japan by the Treaty of Shimonoseki.

The Japanese ruled for the next fifty years, from 1895-1945. While Taiwan was still under Japanese rule, the Republic of China (ROC) was founded on the mainland on 1 January 1912, following the Xin Hai Revolution. In September 1945 following the Japanese surrender in World War II, ROC forces, assisted by small American teams, prepared an amphibious lift into Taiwan to accept the surrender of the Japanese military forces there and take over the administration of Taiwan.

After the end of World War II, the Chinese Civil War resumed between the Chinese Nationalists (Kuomintang), led by Generalissimo Chiang Kai-shek, and the Chinese Communist Party (CCP), led by CCP Chairman Mao Zedong. On 7 December 1949, Chiang evacuated the Nationalist government to Taiwan and made and made Taipei the temporary ROC "wartime capital." Approximately 2 million people, mainly of soldiers, members of the ruling Kuomintang and intellectual and business elites, were evacuated from mainland China to Taiwan, adding to the earlier population of approximately six million. Since losing control of mainland China in 1949, the Kuomintang continued to claim sovereignty over "all of China," which it defined as mainland China (including Tibet), Taiwan (including Penghu), Outer Mongolia, and other minor areas. The People's Republic of China (PRC) claims that Taiwan is Chinese territory and that it replaced the ROC government in 1949, becoming the sole legal government of China. Though it was a founding member of the United Nations, the Republic of China (ROC) now has neither official membership nor observer status in the organization.

China and Taiwan Border Issues

The Taiwan military has approximately 170,000 active duty troops (90,000 Army; 40,000 Navy, including approximately 10,000 marines; 40,000 Air Force) (2021). The Taiwan military is armed mostly with second-hand weapons and equipment provided by the US; Taiwan also has a domestic defense industry capable of building and upgrading a range of weapons systems, including surface naval craft and submarines (2021). The **U.S. Taiwan Relations Act** of April 1979 states that the US shall provide Taiwan with arms of a defensive character and shall maintain the capacity of the US to resist any resort to force or other forms of coercion that would jeopardize the security, or social or economic system, of the people of Taiwan (2021).[12]

(CIA World Factbook, 2022)

The **People's Liberation Army (PLA)** consists of Ground Forces, Navy (PLAN, includes marines and naval aviation), Air Force (PLAAF, includes airborne forces), Rocket Force (strategic missile force), and Strategic Support Force (information, electronic, and cyber warfare, as well as space forces). The Strategic Support Force includes the Space Systems Department, which is responsible for nearly all PLA space operations, including space launch and support, space surveillance, space information support, space telemetry, tracking, and control, and space warfare.[13] (CIA World Factbook)

The **People's Armed Police (PAP)** includes Coast Guard, Border Defense Force, and Internal Security Forces. Forces also include the PLA Reserve Force (2021). The PAP is a paramilitary police component of China's armed forces that is under the command of the Central Military Commission (CMC) and charged with internal security, law enforcement, counterterrorism, and maritime rights protection. In 2018, the Coast Guard was moved from the State Oceanic Administration to the PAP; in 2013, China merged four of its five major maritime law enforcement agencies—the China Marine Surveillance (CMS), Maritime Police, Fishery Law Enforcement (FLE), and Anti-Smuggling Police – into a unified coast guard. (CIA World Factbook)

Information varies, but approximately 2 million total active duty troops (approximately 1 million Ground; 250,000 Navy/Marines; 350-400,000 Air Force; 120,000 Rocket Forces; 150,000-175,000 Strategic Support Forces) make up the PLA and there are an additional estimated 600,000-650,000 People's Armed Police (2021). The PLA is outfitted primarily with a wide mix of older and modern domestically-produced systems heavily influenced by technology derived from other countries. Russia is the top supplier of foreign military equipment since 2010. The Chinese defense-industrial sector is large and capable of producing advanced weapons systems across all military domains; it is the world's second largest arms producer (2021). The PLA is in the midst of a decades-long modernization effort. In 2017, President XI set three developmental goals for the force: becoming 1) a mechanized force with increased information and strategic capabilities by 2020, 2) a fully modernized force by 2035, and 3) a worldwide first-class military by mid-century. (CIA World Factbook, 2022)

The ***Mainland Affairs Council*** (MAC) of Taiwan (ROC) is responsible for relations with the PRC, while the ***Taiwan Affairs Office*** (TAO) of the PRC is responsible for relations with Taiwan. Exchanges are conducted through two corresponding private organizations founded in 1991:

---

12   CIA World Factbook – Taiwan

13   CIA World Factbook – China

- The **Straits Echange Foundation** (**SEF**), a semiofficial organization of the Republic of China (ROC or Taiwan) to handle technical and/or business matters with the PRC. Technically a private organization, SEF is funded by the government and under the supervision of the Mainland Affairs Council, essentially functioning as the *de facto* embassy to the PRC, as a means of avoiding acknowledgement of the PRC's statehood status.

- The **Association for Relations Across the Taiwan Straits** (**ARATS**) is an organization set up by the People's Republic of China (PRC) for handling technical or business matters with Taiwan (ROC). Negotiations with SEF ceased in 1999.

The Republic of China (ROC or Taiwan) and the United States signed the **Sino-American Mutual Defense Treaty** in 1954 (effective from 1955 to 1980), and established the United States Taiwan Defense Command. About 30,000 US troops were stationed in Taiwan, until the United States established diplomatic relations with the PRC in 1979.

National Public Radio (NPR) reported on January 28, 2022, that China's ambassador to the United States issued a warning that the U.S. could face "military conflict" with China over the future status of Taiwan. Qin Gang accused Taiwan of "walking down the road toward independence," and added, "If the Taiwanese authorities, emboldened by the United States, keep going down the road for independence, it most likely will involve China and the United States, the two big countries, in a military conflict." With American eyes focused on a threatened war in Ukraine, U.S. officials and analysts have concerns about Taiwan's ability to defend itself, following 39 Chinese military aircraft flying near Taiwan recently. (National Public Radio, 2022)

The New York Times reported that "25 Chinese fighter jets, bombers and other warplanes flew in menacing formations off the southern end of Taiwan" in a show of military force on China's National Day, on October 1, 2021. The incursions continued into the night and the days that followed and surged to the highest numbers ever on Monday, when 56 warplanes tested Taiwan's beleaguered air defenses. Taiwan's jets scrambled to keep up, while the United States warned China that its "provocative military activity" undermined "regional peace and stability." China did not cower. When a Taiwanese combat air traffic controller radioed one Chinese aircraft, the pilot dismissed the challenge with an obscenity involving the officer's mother. The *Times* report said, "As such confrontations intensify, the balance of power around Taiwan is fundamentally shifting, pushing a decades-long impasse over its future into a dangerous new phase." (The New York Times, 2022)

The BBC reported that in 2021, China ramped up pressure on Taiwan by sending military aircraft into Taiwan's Air Defence Zone, a self-declared area where foreign aircraft are identified, monitored, and controlled in the interests of national security. Taiwan began making data on plane incursions public in 2020 and the numbers of aircraft reported peaked in October 2021, with 56 incursions in a single day. (BBC News, 2022)

Why does this matter? Why is Taiwan important or an international security interest? Much of the world's everyday electronic equipment - from phones to laptops, watches and games consoles - is powered by computer chips made in Taiwan. Taiwan dominates the global production of computer chips, responsible for 65% of chip production. A Chinese takeover in Taiwan could give Beijing some control over one of the world's most important industries.

United States Border Security Issues

While the rest of the world has its border issues, which may result in transnational security threats, the United States shares some of these concerns but also has unique issues involving homeland security and national sovereignty. The Pew Research Center reports that, "Immigrants with past criminal convictions accounted for 74% of all arrests made by U.S. Immigration and Customs Enforcement (ICE) agents in fiscal 2017." The remainder were classified as "non-criminal" arrestees, including 16% with pending criminal charges and 11% with no known criminal convictions or charges. (Pew Research Center, Kristen Bialek, 2022)

The Heritage Foundation reported that while non-citizens constitute only about 7% of the U.S. population, the Justice Department's Bureau of Justice Statistics notes that non-citizens accounted for a staggering 64% (nearly two-thirds) of all federal arrests in 2018. These arrests aren't just for immigration crimes. Non-citizens accounted for 24% of all federal drug arrests, 25% of all federal property arrests, and 28% of all federal fraud arrests. (The Heritage Foundation, Hans A. von Spakovsky, 2019)

In 2018, a quarter of all federal drug arrests took place in the five judicial districts along the U.S.-Mexico border, reflecting the ongoing activities of Mexican drug cartels. More Mexicans than U.S. citizens were arrested on charges of committing federal crimes in 2018. Migrants from Central American countries are also accounting for a larger share of federal arrests, rising from only 1% of arrests in 1998 to 20%. A recent report from the Texas Department of Public Safety revealed that 297,000 non-citizens had been "booked into local Texas jails between June 1, 2011 and July 31, 2019." So these are non-citizens who allegedly committed local crimes, not immigration violations. The report noted that a little more than two-thirds (202,000) of those booked in Texas jails were later confirmed as illegal immigrants by the federal government. According to the Texas report, over the course of their criminal careers those illegal immigrants were charged with committing 494,000 criminal offenses. (The Heritage Foundation, Hans A. von Spakovsky, 2019)

Convictions represent:

- 500 homicides,
- 23,954 assaults
- 3,122 sexual assaults,
- 3,840 sexual offenses,
- 297 kidnappings,
- 2,026 robberies,
- 3,158 weapon charges,
- 8,070 burglaries,
- 14,178 thefts, and
- tens of thousands of drug and obstruction charges.

(The Heritage Foundation, Hans A. von Spakovsky, 2019)

These statistics reveal the very real danger created by illegal border crossings and sanctuary policies. In nine self-declared sanctuary States and numerous sanctuary cities and counties, officials refuse to hand over criminals who are known to be in this country illegally after they have served their state or local sentences. This refusal to cooperate with federal immigration officials suggests that State and local officials supporting the sanctuary movement believe it's better to let these criminals return to their communities rather than being removed from this country. The Texas report is careful to note that it is not claiming "foreign nationals" commit "more crimes than other groups." The Heritage Foundation reports says the Texas report "identifies thousands of crimes that should not have occurred

and thousands of victims that should not have been victimized because the perpetrators should not be here." (The Heritage Foundation, Hans A. von Spakovsky, 2019)

Conclusion

A number of targeted operations and exercises are carried out against specific crimes, including:

- People smuggling
- Human trafficking
- Stolen/lost travel documents, counterfeit documents, and altered documents
- Foreign terrorist fighters
- Drugs, firearms, illicit goods, stolen motor vehicles and stolen vessels
- Smuggling of chemical, explosive, nuclear, biological, and radiological materials.

Land, sea and air borders are essential to maintaining the security of any sovereign nation. Because protecting borders is one of the key functions of government, unsecured borders undermine a public's trust in their government and the rule of law. The three major components of immigration control—deterrence, apprehension and removal—need to be strengthened by Congress and the Executive Branch to restore confidence in government and leadership and to ensure national security and homeland security. (FAIR, 2022)

Department of Homeland Security
U.S. Citizenship and Immigration Services

I-797, Notice of Action

THE UNITED STATES OF AMERICA

RECEIPT NUMBER		CASE TYPE I821 /I-821D
RECEIPT DATE August 29, 2012	PRIORITY DATE	APPLICANT
NOTICE DATE October 12, 2012	PAGE 1 of 1	

Notice Type: Approval Notice
Valid from 10/12/2012 to 10/11/2014

Notice of Deferred Action

Please see the additional information on the back. You will be notified separately about any other cases you filed.
IMMIGRATION & NATURALIZATION SERVICE
TEXAS SERVICE CENTER
P O BOX 851488 - DEPT A
MESQUITE TX 75185-1488
Customer Service Telephone: (800) 375-5283

Form I-797 (Rev. 01/31/05) N

Form I-797 to communicate with applicants/petitioners
or convey an immigration benefit.

References

Phelps, James R., Jeffrey Daily, and Monica Koenigsberg. *Border Security* (Second Edition). Carolina Academic Press; Durham, NC: 2018.

World Health Organization, 2022. About WHO. https://www.who.int/about

World Health Organization, 2022. Coronavirus Disease (COVID-19) Pandemic. https://www.who.int/emergencies/diseases/novel-coronavirus-2019

Department of Homeland Security, 2022. Post-Quantum Cryptography. https://www.dhs.gov/quantum

Cyber Security & Infrastructure Security Agency, 2022. Cyber Security. https://www.cisa.gov/cybersecurity

Office of the Director of National Intelligence, 2022). Cyber Security. https://www.dni.gov/index.php/ncsc-what-we-do/ncsc-cyber-security

UN Chronicles, 2022. Cyber Conflicts and National Security. https://www.un.org/en/chronicle/article/cyberconflicts-and-national-security

CNN Politics, 2021. After 20 years of anti-terror work, CIA gets back to spycraft basics in shift to China. https://www.cnn.com/2021/12/10/politics/cia-shift-china-train-manage-spies/index.html)

AFIO, 2022. John Lenczowski PhD, Former NSC Staffer/Expert, on Chinese Influence Operations in the U.S. https://www.youtube.com/watch?v=i3-8YsvrCEU

Eva Dou, Washington Post, 2022. China warns foreign Olympic athletes against speaking out on politics at Winter Games. https://www.washingtonpost.com/sports/olympics/2022/01/19/china-winter-olympics-politics-speech/

Douglas Main, National Geographic, 2022. Poland's border wall to cut through Europe's last old-growth forest. https://www.nationalgeographic.com/environment/article/polish-belarusian-border-wall-environmental-disaster)

Aljazeera. 2022. Poland begins work on $400m Belarus border wall against refugees. https://www.aljazeera.com/news/2022/1/25/poland-begins-work-on-400m-belarus-border-wall-against-migrants)

Paul Kirby, BBC News, 2022) Why did Russia invade Ukraine and has Putin's war failed? https://www.bbc.com/news/world-europe-56720589

CIA World Factbook, 2022. https://www.cia.gov/the-world-factbook/countries/taiwan/

National Public Radio, Steve Inskeep, 2022. China's ambassador to the U.S. warns of 'military conflict' over Taiwan. https://www.npr.org/2022/01/28/1076246311/chinas-ambassador-to-the-u-s-warns-of-military-conflict-over-taiwan

The New York Times. 'Starting a Fire': U.S. and China Enter Dangerous Territory Over Taiwan, 2022. https://www.nytimes.com/2021/10/09/world/asia/united-states-china-taiwan.html

BBC News, 2022. China and Taiwan: A really simple guide. https://www.bbc.com/news/world-asia-china-59900139

Pew Research Center, Kristen Bialek, 2022. Most immigrants arrested by ICE have prior criminal convictions, a big change from 2009. https://www.pewresearch.org/fact-tank/2018/02/15/most-immigrants-arrested-by-ice-have-prior-criminal-convictions-a-big-change-from-2009/

The Heritage Foundation, Hans A. von Spakovsky, 2019. Crimes by Illegal Immigrants Widespread Across U.S.—Sanctuaries Shouldn't Shield Them. https://www.heritage.org/crime-and-justice/commentary/crimes-illegal-immigrants-widespread-across-us-sanctuaries-shouldnt

FAIR, 2022. https://www.fairus.org/issues/illegal-immigration

The Essence of the Prigozhin Phenomenon

Eugene Alexander Vertlieb

Translated by Dennis T. Faleris

1. The war in Ukraine revealed the defective nature of the feudal-oligarchic system that has built up over the course of three decades in the Russian Federation—a system untenable for a nation, untenable for the majority of its people (more than 20 million living below the poverty line), and untenable for maintaining the sovereign existence of the Russian state itself. Robbed by Chubais-style privatization[1] and disadvantaged by the criminal revolution of the "dashing 90s," the marginal elements of the dispossessed ("the declassed elements"), with their increasing grumbling, have become more and more dangerous for the regime—an anti-Russian establishment based upon exclusion.

2. The Kremlin's political technologists slept through the moment when the people were being excluded from possessing power—a "tectonic dislocation" of social strata. And, like a shift in the earth's crust, a spontaneous revolt of those "hapless" folk broke out, led by Yevgeny Prigozhin[2], the leader of the Private

---

1 Rosalsky, Greg. (2022, March 22). How 'shock therapy' created Russian oligarchs and paved the path for Putin. *Planet Money, NPR.* https://www.npr.org/sections/money/2022/03/22/1087654279/how-shock-therapy-created-russian-oligarchs-and-paved-the-path-for-putin. In the early 1990s, Russian economist Anatoly Borisovich Chubais, was responsible for privatization in Russia. "After the first phase of privatization ended, Chubais...turned to a shady scheme known as 'Loans For Shares'... the richest oligarchs loaned the government billions of dollars in exchange for massive shares of Russia's most valuable state enterprises. When the government defaulted on paying back the loans...the oligarchs would walk away with the keys to Russia's most profitable corporations. In exchange, the government would get the money it needed to pay its bills, privatization would keep moving forward—and, most importantly, the oligarchs would do everything in their power to ensure Yeltsin was reelected."

2 (2023, June 25). Yevgeny Prigozhin: From Putin's chef to rebel in chief. *BBC News online.* After serving 9 years behind bars...Prigozhin set up a chain of hot dog stalls in St. Petersburg...and eventually was able to open expensive restaurants in the city... Putin liked one of Prigozhin's restaurants... [and] started taking foreign guests there. Years later, Prigozhin's catering company...was contracted to supply food to the Kremlin, earning him the nickname "Putin's chef." After Russia's invasion of Ukraine in 2014...a shadowy private military company PMC Wagner] said to be linked to him was first reported to be fighting Ukrainian forces in the eastern Donbas region... Wagner was [also] active across Africa and beyond, invariably performing tasks that furthered the Kremlin's agenda... For years, Prigozhin denied having any links to Wagner...but in September 2022, he said he had set up the group in 2014... He has been accused of being behind so-called 'troll farms' or 'bot factories', which used accounts on social media...to spread pro-Kremlin views. Such efforts were led by the St. Petersburg-based Internet Research Agency best known for meddling in the 2016 US presidential election.

doi: 10.18278/gsis.8.1.11

Military Company (PMC) Wagner. On the night of 24 June 2023, Prigozhin headed to Moscow "seeking the heads" of the country's military leadership. But the campaign, according to him, was not a rebellion and not an attempt to seize power. The group merely disarmed the military; it did not use violence against them. Many local residents took pictures with PMC fighters and greeted them without fear. Wagner fighters opened fire on Russian aviation elements solely for self-defense. On the way to the capital, the Prigozhinites seized all military facilities, including the headquarters of the Southern Military District in Rostov-on-Don. It is paradoxical, but true: Almost passing unhindered through Voronezh, Lipetsk, and Tula oblasts, the battle-hardened rebels suddenly allowed themselves to be persuaded by Belarusian President Alexander Lukashenko to return to their field camps. But there was no doubt about the seriousness of the intentions of the PMC fighters, who had inflicted great damage to the Russian Federation Armed Forces by launching a massive onslaught: Over the course of one day, their unwitting "spoils" amounted to the lives of fifteen combat pilots, one aircraft (a strategic IL-22—an airborne command post), and six helicopters.

3. Analysts guess at the semantic essence of the Prigozhin phenomenon; they consider the coup as having ended in failure. I disagree. I believe that the strategic thought behind Prigozhin's "freedom march"—the first essentially armed rebellion in post-Belovezha Russia[3]—actually was to give the authorities of the Russian Federation a last, pre-revolutionary warning—a flare signaling a directed (at the overthrow of the military leadership) and controlled rebellion, and that, like a signal rocket, illuminated the locations of the various troops, all antagonists in the internal Russian confrontation.

4. It is absurd to deny the obvious internal political crisis in the Russian Federation. This manifestation of the anger stemming from popular dissatisfaction is a harbinger, if not of a cardinal transformation, then of the inevitable fall of a regime, which is deaf and dumb to the aspirations of the common people. The march of freedom (this phrase is true and requires no quotes) awakened the dormant archetypes of the Russian collective unconscious that had been driven underground, and swirled them together with the principles of faith, justice, and honor. From that day forward, there was a guiding beacon for the self-directed movement of the nation towards

---

3   (2016, December 7). History in the Making: The Agreement That Ended the Soviet Union. *The Moscow Times*. The Belovezha Accords ended the Soviet Union and established the Commonwealth of Independent States. https://www.themoscowtimes.com/2016/12/07/history-in-the-making-the-agreement-that-ended-the-soviet-union-a56456

the values which the nouveau riche had partially erased. There are some native banners around which a torn society can come together. Without finding their spiritual tablets, Russians[4] cannot return to the awareness of themselves as a powerful and invincible people. Without the fundamental foundations of victory, Russia is doomed to defeat in the hybrid war centering on Ukraine.

5. Those who steer the post-Belovezha, non-revolutionary Russian Federation are faced with the situation of having to cut off the branch on which they have lived, satisfied and comfortably until now. Their plan is, as Pushkin put it, "to wake up once they are patriots."[5] The top *nomenklatura* officials, cursing their fate and after the fact (since Prigozhin's thunder had already struck!), put on a patriotic mask in quick order. Like cowards tucking their tails between their legs, or becoming shameless from despair, they align

with the people's will—the changing orientation of Russian society's values, which have significantly shifted "from a global liberal plan toward one of justice, honesty, courage, and true front-line brotherhood." The potential of Prigozhin is based on the concept of justice, which is sacred to the Russian consciousness. And this is the key to victory. The Supreme Commander-in-Chief also calls for justice for his "cook": It turns out that billions in the budget were set aside to pay for Wagner jobs. The winners of the rebellion so far look to be the security officials close to Putin. There is no talk of Shoigu's resignation yet, but "you don't change horses in midstream," and the era of justice within the officialdom of the Russian Federation is just beginning.

6. As strategist Alexander Dugin[6] prophetically pointed out back on 8 April 2023, "Our war is a war for justice... It is also directed... against

---

4 Here the author uses "Rusichi" [русичи], an old Russian poetic name for the inhabitants of Kievan Rus. https://ru-m-wikipedia-org.translate.goog/wiki/Русичи?_x_tr_sl=ru&_x_tr_tl=en&_x_tr_hl=en&_x_tr_pto=sc

5 Иванов, Андрей С. (2014, October 26). Путешествие Онегина, расширенная версия. Стихи.ру. [Ivanov, Andrey S. Onegin's journey, an expanded version. *Stikhi.ru.*] Eugene Onegin (1831) is a novel in verse by Russian poet Alexander Pushkin. Onegin, a superfluous "cosmopolitan," seeks to transform, to be someone meaningful. Using Pushkin's phrasing which describes Onegin's hasty patriotism, the author suggests that in the wake of great change, those who now must take actions that will adversely affect their comfortable lives are planning to become (hasty and half-hearted) patriots.

6 Amarasinghe, Punsara. (2020, April 8). Alexander Dugin's Neo Eurasianism in Putin's Russia. *Modern Diplomacy.* https://moderndiplomacy.eu/2020/04/08/alexanders-dugins-neo-eurasianism-in-putins-russia/. "Today [far-right political philosopher Aleksandr Gelyevich Dugin] stands as a stalwart in Russian nationalism...and his Neo-Eurasia project has consolidated many forces against the US and its liberal values... Taking Moscow as an idea standing for the Orthodox creed based on Filofi's 16th century 'Third Rome Doctrine,' the alternative suggested by Dugin invokes the Eurasian nations to drift away from the Atlantists led by the US ad its market civilization."

the injustice that sometimes happens within Russia itself. The war of PMC Wagner is a people's war, one of liberation and cleansing. The war does not accept half-measures, agreements, compromises, or negotiations made behind the backs of fighting heroes."[7] A. Dugin dismisses a priori the inevitable creation in the media of conjectures about the alleged material motives that drove Prigozhin's rebellion: "PMC Wagner is not about money. That has nothing to do with it. PMCs are a military brotherhood, a Russian guard, which Prigozhin assembled from those who responded to the call of the Motherland at a most difficult moment and who went to defend it, being ready to pay any price... For Russians, Prigozhin became the main symbol of victory, determination, heroism, courage, and perseverance. For the enemy, a source of hatred, and, at the same time, fear and horror. It is important to bear in mind that Prigozhin does not simply head the most combat-ready,

victorious, and invincible unit in the Russian Armed Forces; he also gives vent to the feelings, thoughts, demands, and hopes that live in the hearts of people at war, completely and until the end, and remains permanently immersed in its element." The "elite" are clearly slandering Prigozhin, claiming that he is just striving for power and, relying on the people's support, is preparing a black redistribution.[8]

7. The function of Prigozhin is to attack the rotten bosses with an oprichnina spirit of retribution.[9] Prigozhin-Wagnerites—why not see them as "forerunners of a full-fledged oprichnina? After all, in the era of Ivan IV [Ivan the Terrible], the oprichny army was formed in battles in the same way as PMC Wagner, with forces drawn from among the most courageous, daring, desperate, strong, reliable, and active personnel, regardless of pedigree, title, status, rank, or position," as Dugin boldly likens the Wagner

---

7   (2023, 8 April). Александр Дугин: Фактор Евгения Пригожина и тезис справедливости. Четыре Пера. [Aleksandr Dugin: The Evgeny Prigozhin factor and a thesis on justice. *Chetyre Pera.*]   https://4pera.com/news/analytics/aleksandr_dugin_faktor_evgeniya_prigozhina_i_tezis_spravedlivosti/

8   In 1917...the Party of the Socialist Revolutionaries adhered to the theory of 'peasant socialism'.... [Their] slogan 'socialization of agriculture' corresponded to the aspirations of the bulk of the peasantry awaiting a 'black redistribution' of the landlord's land. Black Redistribution [Чёрный передел or Chyornyj Peredel], referred to the long-anticipated seizure and redistribution of all nonpeasant lands. https://www.encyclopedia.com/history/encyclopedias-almanacs-transcripts-and-maps/black-repartition

9   The oprichnina [опричнина] was a state policy implemented by Tsar Ivan the Terrible in Russian between 1565 and 1572. The policy included mass repression of the boyars (Russian aristocrats), including public executions and confiscation of their land and property. It can also refer to the notorious organization of six thousand Oprichniki, the first political police in the history of Russia. https://en.wikipedia.org/wiki/Oprichnina

troops to those of the oprichnina.[10]

8. A meritocracy has replaced the system of tsarist times that valued loyalty and courtier skills. Now there is the power of the new passionaries—of the effective war elite or of crisis managers. Wagner principles are being established in all areas: Those who are the most effective in coping with the most difficult task given them are the ones who dominate. A rotation of the elites is ultimately required by the war. What causes real horror for the elites who are old and who have lost the capacity to act, moreover, is that they have been cut off from the matrix in the West. Yevgeny Prigozhin has designated the most important direction vector in which Russia will have to move under any conditions and in any circumstances if it is to avoid total strategic defeat.

9. The abundance of versions in the interpretation of these single-day rebellious events suggests that there are more questions than answers.

"Why did Vladimir Putin back down—or did he? Why did Prigozhin deploy the columns? And what about Belarus in general, and what will happen to PMC Wagner now? How will all of this affect the war in Ukraine?" The truth gets buried in an avalanche of "fake news"—a symbiosis of the real truth and what we wish the truth to be—generated by a neural network. Since all the actors involved in this coup are Heroes of Russia recipients (Prigozhin has already been awarded the Hero three times—the Hero of Russia, the Hero of the Donetsk People's Republic, and the Hero of the Luhansk People's Republic) ,[11] the people sarcastically reacted to this force majeure as if were a "friendly get-together to strike some sort of deal." They say "the defenders of Russia decided to seize power in Russia. Therefore, other defenders of Russia rushed in to kill the first defenders of Russia, but then they themselves got killed. And Hero of Russia Prigozhin went to kill Hero of Russia Shoigu.[12] Because of this,

10 (2023, April 8). Дугин: Пригожин и ЧВК «Вагнер»—это новая опричнина в России. *Капитал Страны, Федеральное Интернет издание* [Dugin: Prigozhin and PMC 'Wagner'—a new Oprichnina in Russia. *Kapital Strany, Federal Internet Edition*]. https://kapital-rus.ru/news/395084-dugin_prigojin_i_chvk_vagner__eto_novaya_oprichnina_v_rossii/

11 Goncharenko, Roman. (2022, February 23). Donetsk and Luhansk: A tale of creeping occupation. *DW In Focus: Politics-Ukraine.* "The self-proclaimed People's Republics of Donetsk and Luhansk were formed in spring 2014, following Ukraine's pro-Western opposition protests and a change of leadership in Kyiv... Since the conflict broke out, millions of people have left the separatist areas. A majority of the civilians fled to Ukraine, hundreds of thousands to Russia. https://www.dw.com/en/donetsk-and-luhansk-in-ukraine-a-creeping-process-of-occupation/a-60878068

12 Mandraud, Isabelle. (2023, June 30). Who is the real Sergei Shoigu, Russia's defense minister? *Le Monde.* Sergey Shoigu, Minister of Defense, Russian Federation, since 18 May 2018 "crashed suddenly with the invasion of Ukraine by Russian troops in February 2022. The mounting failures in the face of Ukrainian resistance, the abuses committed by his troops, the excessive cost

Hero of Russia Kadyrov[13] went to kill Hero of Russia Prigozhin. And because of this, Hero of Russia Bortnikov[14] opened a criminal case against Hero of Russia Prigozhin, but immediately closed it[15] because the most important Hero of Russia, Putin, at first guaranteed that the traitors would be punished, but then guaranteed that they would not. And there is no one who could not care less about those very first defenders who were killed. The main thing is that all the defenders and heroes can continue to be proud of what heroic defenders Russia has. The rebellion took place in a warm and friendly atmosphere."[16]

10. Of course, deadly internecine strife among generals during a war is fraught with betrayal and calls for the death penalty for what has taken place. "This is a stab in the back. The same blow was dealt in 1917: Russians were killing Russians and the West got the benefit. What we are facing is betrayal. Any internal threat is deadly. Prigozhin is pushing the country into anarchy and confusion. As a result, defeat will follow." This was how Putin assessed the attempted military uprising in the country.[17] And if this uprising was a spontaneous reaction from a true patriot who is close to Putin and who was the only one who heard the SOS of the Russian world crying to itself, and who, like performing an urgent surgery, took an extraordinary measure to save the Fatherland on the edge of the

---

of troop losses all marked him as a target for Yevgeny Prigozhin." https://www.lemonde.fr/en/m-le-mag/article/2023/06/30/who-is-the-real-sergei-shoigu-russia-s-defense-minister_6039898_117.html

13 Osborn, Andrew. (2023, June 24). Chechen leader offers to help put down Wagner mutiny. *Reuters*. "Chechen leader Ramzan Kadyrov said...his forces were ready to help put down a mutiny by Wagner mercenary chief Yevgeny Prigozhin and to use harsh methods if necessary. He said that Chechen units were moving towards the 'zones of tension' and would act to 'preserve Russia's units and defend its statehood.'" https://www.reuters.com/world/europe/chechen-leader-says-his-forces-are-ready-help-put-down-wagner-mutiny-2023-06-24/

14 The Federal Security Service (FSB) has been headed by Alexander Vasilyevich Bortnikov since 2008.

15 The FSB case was opened on the 23d of June. On 27 June, however, "Russian authorities said they have closed a criminal investigation into the uprising and are pressing no charges against Prigozhin or his troops after the negotiated deal... The FSB said its investigation found that those involved... 'ceased activities directed at committing the crime,' so the case would not be pursued." https://www.pbs.org/newshour/world/russia-drops-charges-against-prigozhin-and-others-who-took-part-in-aborted-armed-rebellion

16 (2023, June 26). Вчера защитники России решили захватить власть в России [Yesterday defenders of Russian decided to seize power in Russia]. *Live Journal broadcast*. https://zermd.livejournal.com/1435807.html?

17 Mezdrikov, Evgeny. (2023, June 24). Владимир Путин назвал мятеж ударом в спину. Ведомости. [Vladimir Putin called the rebellion a stab in the back. *Vedomosti*.] https://www.vedomosti.ru/politics/articles/2023/06/24/982107-chto-skazal-putin-o-myatezhe-prigozhina.

abyss, then shouldn't it be considered that Prigozhin's military-patriotic demarche was a self-detonation by those refusing to surrender to the ruling oligarchy-pakhanate of the sovereign-national part of the Russian imperial conglomerate? I repeat, Prigozhin and his oprichniki absolutely did not prepare a coup d'état—yet, that is—thereby naively allowing the leaders themselves to change the course of the country to a diametrically opposite one. He only sent a message to the military-political "elite" about the further intolerance of the people to the existing monstrous inequality among fellow citizens of the Russian Federation. With Prigozhin's protest movement and the harmful resistance of the sympathizer "fifth column," the specter of a civil war loomed.

11. Yevgeny Prigozhin violated the balance of clan interests backed by President Vladimir Putin himself who acts as referee. Political forecaster Valery Morozov states that Putin has managed to create a system that ensures the power of one group while maintaining a balance of interests among the other clans of the ruling elite. However, "in conditions of war, the army, special services, and the military-industrial complex are not tools, but are the most active actors in the political struggle."[18] The Prigozhin phenomenon has weakened Putin's power and has objectively furthered change within the ruling regime, the political and economic foundation for which will be determined or created by the clans representing the Russian armed forces and special services that won the struggle. "In conditions of war, power is concentrated in the hands of those who, on one hand, can most severely subjugate, force, and destroy rivals, and on the other, can take into account and reflect the interests and moods of the masses to the maximum extent, and therefore can ensure the protection and stability of the state in the event that the people are drawn into the political struggle—and who can as well ensure victory in the war. In today's Russia, it is the army and special services that have become such forces in the state, and they seek to push the state bureaucracy—including Ministry of Defense functionaries—away from the center of the state's power. And this is precisely what Prigozhin was demanding. Maybe that is why they didn't punish him for all the turmoil? After all, he is on the track that the country needs, but just a little "ahead of the locomotive." Publicist Maxim Kalashnikov assures that "Prigozhin's political project is being supervised from the very top ... a low-tech version (a lesser imitation) of Röhm's

---

18 Морозов, Валерий. (2023, July 6). Военно-политический кризис в России. Причины и факторы влияния. Личный Блог. [The Military-Political Crisis in Russia. Reasons and Factors for its Appearance. *Personal Blog.*] https://valerymorozov.com/news/3502

assault detachments in Germany in the 1920s is being shaped."[19]

12. According to the conclusions of V. Morozov, the clan choosing S. Shoigu as V. Putin's successor was behind the idea of conducting a lightning-fast special operation which was supposed to repeat the operational success of the annexation of Crimea—to become a victory parade for the Putin-Shoigu pair. But the opposing clan prevented Shoigu from emerging from the Special Military Operation (SMO) blitzkrieg as the winner and therefore worthy of "the succession to the throne." The Armed Forces of the Russian Federation, having immediately destroyed Ukrainian aviation, were already positioned near Kyiv. But the SMO suddenly stalled and turned into a no-victor positional war. Thus, Shoigu's chances were reduced to the minimum. And Putin apparently will not have to leave power in the re-election year 2024, but will continue to look for a suitable figure instead of Shoigu. And to that point, the fact that the Western establishment considered

him "the most reliable contact that can be trusted" played a negative role, removing him from the most likely contenders for the country's highest post. But at the same time, "it is necessary to continue to put pressure on Putin from the outside and from the inside. He is able to make peace, which each side will claim as a victory, but in reality, will be the defeat of Russia. The path to such a peace is through escalation, through applying pressure on Russia, through raising the stakes and increasing the threats until the moment when the military—first and foremost the generals of the Ministry of Defense—and the special services understand that Russia cannot withstand it, and force Putin to make peace. Pro-Western groups will give them support."[20] At this point [when the article was published], eleven days remained before Prigozhin's "rebellion."

13. Let's summarize the analyses of the military campaign results, taking into account the recent proposal of the Germans concerning the division of Ukraine.[21] At present, PMC

19 (2023, March 28). Тезисы о Пригоожине и 12 пунктах [Theses about Prigozhin and 12 points]. *LiveJournal*. Maksim Aleksandrovich Kalashnikov (real name Vladimir Aleksandrovich Kucherenko) is a Russian journalist as well as a social and political activist, writer-futurologist, and publicist. https://m-kalashnikov.livejournal.com/4378164.html

20 Morozov, Valery. (2023, 15 June). Запад о Путине, Кремле и перспективах войны на Украине. Личный Блог. [The West on Putin, the Kremlin and the Prospects of the War in Ukraine. *Personal Blog*] http://valerymorozov.com/news/3481

21 See other examples, including: (2022, November 10). Belgian senator: It is in Europe's interest to abandon Zelensky's hard line. *NewsFront*. "Reporting stemming from an interview in the French publication *Causeur* indicated that as early as November 2022, Belgian senator Alain Detex had called upon Europe to initiate negotiations with Russia on the 'partitioning' of Ukraine." https://en.news-front.info/2022/11/10/belgian-senator-it-is-in-europes-interest-to-abandon-zelenskys-

Wagner camps are deployed almost on the border of Belarus and not far from Kyiv. This could mean that they would be able to storm Kiev, and unlike Bakhmut (Artemovsk), do it much faster since the Armed Forces of Ukraine no longer have well-trained, combat-ready formations (although Turkey reinforced the Armed Forces of Ukraine with Azov commanders). An imminent division of Ukraine cannot be ruled out, one in which Belarus could also participate, taking for itself the northern border regions up to Kyiv—and perhaps even Kiev as a nice bonus. The rest would be divided up by Poland and Russia. After all, Polish officials have already been allowed to hold leadership positions within Ukrainian power structures. And this means that the question of the territorial integrity of Ukraine was not raised even back then.

14. The West, in this case, loses not only the Ukrainian military campaign, but the dominance it has enjoyed for 500 years. Consequently, the strategic direction it has taken, which demonized and harassed Vladimir Putin along with his country, will have been an erroneous calculation. (I have written about this in my articles for many years.) Now, when not only in the expert community have they woken up to "a discourse about the exhaustion of the old model of the country's development," but on their own, the people have come to recognize the thoroughly rotten state structure that exists, "it's too late to drink Borjomi if the kidneys have failed,"[22] as they say. Economist Mikhail Khazin[23] voiced on his Telegram channel an extremely interesting assumption that Vladimir Putin in the very near future may choose the path of socialist development as the main strategy for Russia's exit from a crisis which the liberals had led him into. Khazin further explained that the head of state had recently given a clear signal for adopting this option in St. Petersburg, where he hosted a naval parade under the flag of the Soviet Union. In addition,

---

hard-line/ Other reporting points to "fake news" being disseminated ostensibly by Russia that which includes claims of ambitions by a number of countries to divide up Ukraine, yet which reports are reliable remains unclear: "...Since the annexation of Crimea by Russia in 2014, signs have emerged to support these [pro-partition] claims... In March 2022, articles appeared in the Russian media accompanied by a map pretending to show the plans of Poland, Hungary and Romania to invade Ukrainian territories...provided by former deputy of the Ukrainian Supreme Parliament Illia Volodymyrovych Kiva...who was suspended from the parliament for his pro-Russian statements after the start of the war." Nikolova, Vanessa. (2023, February 24). The disinformation about the partition of Ukraine. *Factcheck.bg.* https://factcheck.bg/en/the-disinformation-about-the-partition-of-ukraine/

22 An English would be "It's too late to close the stable door after the horse has bolted."

23 Mikhail Leonidovich Khazin is a famous Russian economist. Among other positions, he is a member of the High Council of the International "Eurasian Movement." https://cratology.org/mikhail-khazin-biography/

Putin has repeatedly expressed positive assessments overall of a socialist order.[24]

15. Another argument in favor of a possible drift towards socialism in the economy is the upcoming shift by Chairman of the Government of the Russian Federation Mikhail Mishustin of Russian industry to a socialist footing along with a reorganization of the state planning system and the restoration of Soviet centralized management for all the most important enterprises.[25] In connection with this, it is possible that Putin could be expected to announce a change in the model of state development from a liberal capitalist model to conservative socialist one. The president of the Russian Federation is expected to publicly repent for erring by following a liberal and market approach. Even a computer would break down if it tried to decode Russian logic. How does it go? No matter what they build, it still winds up looking like the Soviet Union. What you fought for might wind up being your undoing.[26]

---

24   (2023, January 30). Хазин предсказал неожиданный ход Путина [Khazin predicted Putin's unexpected move]. *Deita.ru*. https://www.if24.ru/hazin-predskazal-neozhidannyj-hod-putina/

25   Trickett, Nick. (2022, May 16). Russia's Brittle Wartime Economy. *Riddle*. "Russian Prime Minister Mikhail Vladimirovich Mishustin is aimed at 'fool-proofing' Putinism through the creation of a central coordination center within the government... The appearance of neo-Soviet approaches to strengthening executive control over state organs points to a foundational problem: the regime can scarcely mobilize the economy effectively, never mind debates about whether full military mobilization will take place." https://ridl.io/russia-rsquo-s-brittle-war time-economy/

26   This Russian saying is about a situation when, after having fought for something and achieved it with great difficulty, the result becomes the cause of new troubles—perhaps, ironically, "We've reaped what we've sown." https://verymuchrussian-com.translate.goog/russian-proverbs/za-chto-borolis-na-to-i-naporolis/?_x_tr_sl=ru&_x_tr_tl=en&_x_tr_hl=en&_x_tr_pto=sc

Dr. Eugene A. Vertlieb is a Russian-born dissident holding U.S. citizenship and currently living in France. He received a B.A. at Leningrad (Sankt-Peterburg) State University, a PhD at the University of North Carolina, and completed a postdoctoral internship at the Russian Academy of Public Administration in the Russian Federation. Dr. Vertlieb has held a variety of positions including Professor at the Marshall European Center for Security Studies, Germany, and at the Institute of International Relations (U.S.). He is an author of several books and articles on geopolitics and comparative cultural studies. Dr. Vertlieb has also served as an independent political adviser to the Committee on International Affairs and International Relations, Kyrgyzstan. He was nominated for the Chingiz Aimatov Gold Medal for his contribution in formulating key philosophical elements reflecting Kyrgyz ethnicity for inclusion in the "Concept of National Security for Kyrgyzstan." Additionally, for his active and fruitful involvement in furthering military-diplomatic cooperation between the United States and Mongolia, he was awarded a certificate of Honor of the Mongolian Ministry of Defense signed by the Mongolian General Staff Chief, General Tsevegsuren Togoo. Most recently, Dr. Vertlieb received a token of gratitude for fruitful international cooperation from the Research Institute of the German Council of Foreign Relations and the title of "Knight of Science and the Arts" from the Presidium of the Russian Academy of Natural Sciences. He is currently President of the International Institute for Strategic Assessments and Conflict Management (IISACM-France); Executive Director of the Western Policy Forecasting Department for Slavic Europe, (Munich, Germany); Executive Member of the Lisbon-Vladivostok Initiative (France), and Head of the Department of International Relations of the world Christian organization "Blagovest Media International" (Brussels).

Dennis T. Faleris received a B.S. from the University of Michigan and a Master's degree in Russian Linguistics from Georgetown University. For more than thirty-five years, Mr. Faleris worked as an instructor, translator, senior intelligence analyst, and intelligence production manager at the National Security Agency. His career centered around Soviet/Russian military issues as well as a variety of transnational issues. He currently resides in Annapolis, Maryland, with his wife, Kathleen.

СУЩНОСТЬ ФЕНОМЕНА ПРИГОЖИНА

Д-р Евгений Александрович Вертлиб

1. Войной на Украине вскрылась дефективность отстроенной за три десятилетия в РФ феодально-олигархатной нежизнеспособной для нации системы -- большинству народа (более 20 миллионов за чертой бедности) и суверенному существованию самого российского государства. Обворованные приватизацией по-чубайсу и обездоленные криминальной революцией «лихих 90-х», маргиналы люмпенизации («деклассированные элементы») своим нарастающим ропотом становились всё более опасными для режима антирусского истеблишмента отторжения.

2. Политтехнологи Кремля проспали момент отторжения народа от власти -- «тектонической дислокации» страт в социуме. И, как сдвиг земной коры, вспыхнул стихийный бунт «подзаборного» люда во главе с Евгением Пригожиным -- предводителем военизированного формирования «Вагнер». В ночь на 24 июня 2023 года Пригожин пошёл на Москву «за головами» военного руководства страны. Но поход, по его утверждению, не бунт и не захват власти. Они только разоружали военных, но не применяли к ним насилия.

Многие местные жители фотографировались с бойцами ЧВК и без страха приветствовали их. Огонь по российской авиации бойцы «Вагнера» открывали исключительно в целях самообороны. На пути к столице пригожинцы захватили все военные объекты, включая штаб Южного военного округа в Ростове-на-Дону. Парадоксально, но факт: почти беспрепятственно пройдя Воронежскую, Липецкую, Тульскую области, закалённые в боях повстанцы вдруг дали себя уговорить белорусскому президенту Александру Лукашенко вернуться в полевые лагеря. А ведь не вызывала сомнений серьёзность намерений бойцов ЧВК, массированным натиском нанёсшим большой урон ВС РФ: их невольные «трофеи» ЗА ДЕНЬ составили жизни пятнадцати боевых лётчиков, один самолёт (стратегический ИЛ-22 – летающий командный пункт) и шесть вертолётов.

3. Аналитики гадают над смысловой сущностью пригожинского феномена, считая этот переворот закончившемся неудачей. Не согласен. Полагаю: стратегический смысл «марша свободы» Пригожина – первого по сути

вооружённого мятежа в постбеловежской России – и состоял в том, чтобы сделать властям РФ последнее, предреволюционное предупреждение -- вспышкой направленного (на свержение военной верхушки) и управляемого бунта, как сигнальной ракетой, осветив дислокацию размежевания сил внутригосударственного противоборства антагонистов.

4. Нелепо отрицать очевидный внутриполитический кризис в РФ. Сей знак гнева народного негодования – предвестник если не кардинальной трансформации, то неминуемого падения режима, глухонемого к чаяниям простолюда. Марш свободы (истинный: без кавычек) взвихрил загнанные в подполье дремлющие архетипы русского коллективного бессознательного, с постулатами веры, справедливости, чести. Отныне есть путеводный маяк для самодвижения нации к своим полустёртым нуворишами ценностям. Есть вокруг каких родных стягов консолидироваться разорванному обществу. Без обретения своих духовных скрижалей русичам не вернуться к осознанию себя могучим и непобедимым народом. Без фундаментальных основ победы Россия обречена на поражение в гибридной войне вокруг Украины.

5. Рулящие постбеловежско-недоворотной РФ поставлены перед фактом рубить сук, на котором им уютно и сытно жилось доселе. Повестка дня для них «пушкинская»: проснуться раз им патриотами. Высшие номенклатурщики, проклиная судьбу, постфактумно (гром пригожинский ведь грянул!) в срочном порядке обретают патриотическую личину. Трусливо поджав хвосты, или наглея от безысходности, они липнут к народной воле – меняющимся ценностным ориентирам общества, существенно сместившимся «от глобального либерального проекта в сторону справедливости, честности, мужества и истинного фронтового братства». Потенциал Е. Пригожина зиждется на святом для русского сознания понятии справедливость. А это залог побед. К справедливости в отношении своего «повара» апеллирует и Верховный главнокомандующий: оказывается, что на работу «Вагнера» выделяли миллиарды бюджетных денег. Победителями по итогам мятежа пока выглядят близкие к Путину силовики. Об отставке Шойгу пока не идёт речи. Но «коней на переправе не меняют», да и эра справедливости в официозе РФ только начинается.

6. Как провидчески углядел ещё 8 апреля 2023 года стратег Александр Дугин, «наша война

- война за справедливость... Она обращена... и против той несправедливости, которая подчас творится внутри самой России. Война ЧВК "Вагнера" - народная, освободительная, очистительная. Она не приемлет полумер, договорённостей, компромиссов, переговоров за спиной сражающихся героев». А.Дугин отметает априори и предстоящее муссирование в СМИ догадок о якобы материальной подоплёке мотивации мятежа Пригожина: «ЧВК "Вагнера" не про деньги. Они здесь ни при чём. ЧВК - боевое братство, русская гвардия, которую Пригожин собрал из тех, кто откликнулся на зов Родины в самый трудный момент и пошёл защищать её, будучи готовым заплатить любую цену... Для русских Пригожин стал главным символом победы, решительности, героизма, мужества и стойкости. Для врага - источником ненависти, а одновременно страха и ужаса. Важно, что Пригожин не просто возглавляет самое боеспособное, победоносное и необоримое подразделение Вооружённых сил России, но даёт выход чувствам, мыслям, требованиям и надеждам, которые живут в сердцах людей войны, полностью и до конца, необратимо погруженных в её стихию». «Элита» явно клевещет на Пригожина, что якобы тот просто рвётся к власти и, опираясь на народ, готовит чёрный передел.

7. Функция Пригожина – дохнуть на прогнившее начальство духом возмездия опричнины. Пригожинцы-вагнеровцы -- чем не «предтечи полноценной о-причнины? Ведь в эпоху Ивана IV опричное войско формировалось в боях так же, как ЧВК "Вагнера", из числа самых мужественных, смелых, отчаянных, крепких, надёжных, деятельных - независимо от родословной, титула, статуса, чина, положения"» - смело уподобляет Дугин одно другому.

8. На смену лояльности с царедворческими навыками идёт меритократия -- власть новых пассионариев: эффективной элиты войны, или кризис-менеджеров. Устанавливаются вагнер-принципы во всех областях: доминируют те, кто наиболее эффективно справляются с труднейшей поставленной задачей. Ротация элит ультимативно требуется войной. Что вызывает настоящий ужас для элит старых и утративших дееспособность, к тому же отрезанных от матрицы на Западе. Евгений Пригожин обозначил важнейший вектор того направления, в котором России придётся двигаться при любых условиях и в любых обстоятельствах – чтобы избежать тотального стратегического поражения.

9. Обилие версий в толковании этих однодневных мятежных

событий говорит о том, что о-стаётся больше вопросов, чем ответов. «Почему Владимир Путин пошёл на попятную – и пошёл ли? Почему развернул колонны Пригожин? Причём тут вообще Беларусь и что теперь будет с ЧВК «Вагнер»? Как всё это повлияет на войну в Украине?» Истина вязнет в лавине «фейк news» -- этом симбиозе подлинной правды и желаемого, сгенерированного нейросетью. Поскольку все акторы этого путча Герои России (в ходе обысков стало известно, что Пригожин является аж трижды Героем: России, ДНР и ЛНР), то народ саркастично отреагировал на сей форс-мажор -- как на «договорняковый межсобойчик»: мол «защитники России решили захватить власть в России. Поэтому другие защитники России полетели убивать первых защитников России, но те их сами убили. А Герой России Пригожин ехал убивать Героя России Шойгу. Из-за этого Герой России Кадыров ехал убивать Героя России Пригожина. А ещё из-за этого Герой России Борт-ников возбудил дело против Героя России Пригожина, но тут же закрыл его. Потому что самый главный герой России Путин сначала гарантировал, что предатели понесут нака-зание, а потом гарантировал, что не понесут. А на тех самых первых защитников, которых убили, всем наплевать. Главное,

что все защитники и герои, и можно продолжать гордиться тем, какие у России героические защитники. Мятеж прошёл в тё-плой и дружеской обстановке».

10. Конечно, смертельная междоу-собица в генералитете во время войны чревата предательством, с расстрельным наказанием за содеянное. «Это удар в спину, тот же удар был нанесён в 1917 году, россияне убивали россиян, а выгоду получил Запад. То, с чем мы столкнулись, — это пре-дательство. Любая внутренняя угроза — смертельная. **Приго-жин толкает страну в анархию** и смуту. в результате последует поражение» —оценка Путиным попытки военного восстания в стране. А если это произвольная реакция подлинного патриота из приближенных к Путину, единственно услышавшего SOS вопиющего про себя русского мира -- хирургически неотлож-ная экстраординарная мера по спасению Отечества на краю пропасти?! Тогда не следует ли считать, что пригожинский воен-но-патриотический демарш -- это самодетонация несдавшейся властвующему олигархат-паха-нату суверенно-национальной части российского имперского конгломерата? Повторяю, на-верняка Пригожин со своими опричниками и не готовил ПОКА госпереворота, наивно допуская самими верхами смену курса страны на диаметрально

противоположный. Он лишь послал военно-политической «элите» месседж о дальнейшей нетерпимости народа к сложившемуся чудовищному неравенству между согражданами РФ. С протестным движением Пригожина и вредоносным сопротивлением «пятой колонны» -- замаячил призрак гражданской войны.

11. Евгением Пригожиным нарушен баланс клановых интересов, обеспечиваемый самим президентом Владимиром Путиным в качестве рефери. Политпрогнозист Валерий Морозов констатирует, что Путину удалось создать систему, обеспечивающую власть одной группировки при сохранение баланса интересов других кланов правящей элиты. Однако «в условиях войны армия, спецслужбы и военно-промышленный комплекс являются не инструментами, а наиболее активными акторами политической борьбы». Феномен Пригожина ослабил власть Путина, и объективно способствует смене правящего режима, политический и экономический фундамент для которого будет определён и создан победившими в борьбе кланами, представляющими вооружённые силы и спецслужбы России. «В условиях войны власть концентрируется в руках тех, кто, с одной стороны, может наиболее

жёстко подчинять, заставлять и уничтожать соперников, а с другой стороны, кто максимально учитывает, отражает интересы и настроения народных масс, следовательно, может обеспечить защиту, стабильность государства в случае вовлечения народа в политическую борьбу и обеспечить победу в войне. В нынешней России именно армия и спецслужбы стали такими силами в государстве, и они стремятся оттеснить государственную бюрократию от центра власти в государстве, в том числе чиновников в Министерстве обороны». Но ведь именно этого и требовал Пригожин. Может быть, потому и не покарали его за смуту? Ведь он в колее нужного стране, но лишь чуть «впереди паровоза». Публицист Максим Калашников заверяет, что «политпроект Е.Пригожина курируется с самого верха»: «лепится низкотехнологичный симулякр (прпсу-имитацию) штурмовых отрядов Рёма в Германии 1920-х».

12. По выводам В.Морозова, клан выбора С.Шойгу преемником В.Путина стоял за идеей проведения молниеносной спецоперации, которая должна была повторить оперативный успех по присоединению Крыма -- стать парадом победы спарки Путин-Шойгу. Но противодействующий клан помешал Шойгу выйти из блицкрига СВО

победителем и, следовательно, достойным «престолонаследия». ВС РФ, сходу уничтожив украинскую авиацию, уже стояли у Киева. Но СВО внезапно забуксовало, превращаясь в ничейную позиционную войну. Тем самым шансы Шойгу минимизировались. И Путину, видимо, придётся не уходить из власти в перевыборном 2024 году, а продолжать искать подходящую фигуру вместо Шойгу. Кстати, негативную роль по выводу Шойгу из вероятнейших претендентов на высший в стране пост сыграл тот факт, что западный истеблишмент считал его «наиболее надёжным контактом, которому можно доверять». Но при этом продолжать «давить на Путина надо извне и изнутри. Он способен пойти на мир, который каждая из сторон представит как победу, но в реальности, это будет поражением России. Путь к такому миру — через эскалацию, давление на Россию, повышение ставок и угроз до момента, когда военные, прежде всего, генералитет министерства обороны, и спецслужбы поймут, что Россия не может выдержать, и заставят Путина пойти на мир. Прозападные группы окажут им поддержку.» [(Запад о Путине, Кремле и перспективах войны на Украине » Валерий Морозов (valerymorozov.com)]. До «бунта» Пригожина оставалось одиннадцать дней.

13. Суммируем аналитический зондаж результатов похода, с учётом недавнего предложения немцев о разделе Украины. Сейчас лагеря ЧВК дислоцируются почти на границе Белоруссии и недалеко от Киева. это может означать, что они смогут штурмовать Киев, и в отличие от Бахмута (Артёмовска), сделать это гораздо быстрее, поскольку у ВСУ уже не осталось хорошо подготовленных боеспособных соединений (правда, Турция усилила ВСУ командирами «Азова»). Не исключён скорый раздел Украины, в котором может принять участие и Белоруссия, забрав себе северные приграничные районы, вплоть до Киева. А возможно и с Киевом в качестве приятного бонуса. Остальное поделят Польша и Россия. Ведь польским чиновникам уже разрешали занимать управляющие должности в украинских органах власти. А это значит, что вопрос о территориальной целостности Украины не стоял уже тогда.

14. Запад, в таком случае, проигрывает не только украинскую военную кампанию, но и 500-летние своё доминирование. Следовательно, стратегическая установка на демонизацию и третирование В.В.Путина вместе с его страной – ошибочный расчёт. Об этом я многолетно писал в своих статьях. Теперь же, когда не только в эксперт-

ной среде очнулись «дискурсом об исчерпании старой модели развития страны», но и народ самоходом попёр на основательно подгнившую госконструкцию, как говорится, «поздно пить боржоми, если отказали почки». Экономист Михаил Хазин озвучил в своём Telegram-канале крайне интересное предположение о том, что Владимир Путин уже в самое ближайшее время может выбрать в качестве основной стратегии выхода России из кризиса, в который завели стану либералы, путь социалистического развития. Эксперт пояснил, что явный сигнал к этому выбору глава государства недавно сделал в Петербурге, где он принимал морской парад под флагом Советского Союза. Кроме того, Путин неоднократно высказывал свои комплементарные оценки в целом самому социалистическому укладу как таковому.

15. Ещё один аргумент в пользу дрейфа в сторону социализма в экономике — это готовящийся перевод председателем правительства РФ Михаилом Мишустиным промышленности на социалистические рельсы, с реорганизацией системы государственного планирования и восстановлением советского централизованного управления для всех самых важных предприятий. В этой связи, не исключено, что от Путина можно ожидать объявления о смене модели развития государства с либеральной и капиталистической на консервативно-социалистическую. От президента РФ ожидается публичное раскаяние за ошибочность следования либеральному и рыночному курсу. От раскодировки русской логики даже вычислительная машина ломается. Как так: что бы ни строили – всё равно в итоге Советский Союз? За что боролись – на то и напоролись.

Евгений Александрович Вертлиб/Dr. Eugene Alexander Vertlieb президент Международного института стратегических оценок и управления конфликтами (МИСОУК-Франция); ответственный редактор отдела прогнозирования политики Запада «Славянской Европы» (Мюнхен); экзекьютив член Инициативы «Лиссабон-Владивосток» (Париж).

Black Swan in Intelligence

Romeo-Ionuț Repez, PhD

ROMEOIONUTM@gmail.com

Introduction

The objective of this paper is to present in a possible scenario the concept of the 'black swan' applied to the field of intelligence and where it can occur, so that analyst intelligence officers and operative intelligence officers be more aware of possible human errors that can turn into a 'black swan' for everyone else within the secret service and for the whole national security system.

The world of secret services is 'a game' and a struggle between the minds of human beings. The intelligence and creativity of human beings are constantly tested. The 'intellectual' field of secret services is known as 'intelligence'. But what means 'intelligence'? The concept 'intelligence' refers to '*the product resulting from collection, processing, integration, evaluation, analysis, and interpretation of available information concerning foreign nations, hostile or potentially hostile forces or elements, or areas of actual or operations*' (DOD Dictionary 2020, 107). The concept of '*intelligence*' is always related to the concept of '*counterintelligence*'. The concept 'counterintelligence' refers to '*information gathered and activities conducted to identify, deceive, exploit, disrupt, or protect against espionage, other intelligence activities,*

sabotage, or assassinations conducted for or on behalf of foreign powers, organizations or persons or their agents, or international terrorist organizations or activities' (DOD Dictionary 2020, 52).

For the efficient and effective interconnection of the national system of secret services, democratic states have created the National Intelligence Community. The National Intelligence Community has its activities based on information. These information are obtained through various and complex methods, means, and procedures either by intelligence officers from operative or from HUMITN, SIGINT, IMINT, MASINT, OSINT, GEOINT, TECHINT and CYBERINT. Many of those information are processed by analysts in order to give the most effective solutions to a certain situation or event. Then based on this it can be created strategies regarding the directions of a secret service conducting its existence, its vision and its missions.

Secret services deliver to the political decision makers the information and solutions to generate the politics and strategies to govern the cities and the country. The main problem is that the intelligence provided by secret services are not entirely taken into account by political decision makers. Because of this the governance of the cities or the governance of the country can be jeop-

 doi: 10.18278/gsis.8.1.12

ardized by its own leaders. This is an inner issue that can be exploited by secret services.

Another problem is that some of the secret services provides intelligence tailored to their own needs and deliver them to political decision makers. This sensitive issue can raise the question: are political decision makers still the guarantors of the rule of law and the good governance? Because a high sensitive issue is using the secret services' intelligence for political purposes (Hans Born and Aidan Wills, 2012, 155).

The secret services' primordial aim is to safeguard its own country by any known methods, means, and procedures. The control and overseeing of the secret services is carried out by institutions that are under the power of political decision makers. So here is the game table between secret services and politics. The main stakes between secret services and politics are power and money. To obtain power of influence one over another is to have the best human assets to occupy key positions either in the hierarchy of secret services and/or in the hierarchy of politics (Parliament, Government, secret service directors and other public institutions). Money is the other stake that both categories wants to control much more of them.

The concept of black swan in intelligence

Further, I present the concept as it was created by Nassim Nicholas Taleb, with my empirical obser-

vations regarding its characteristics in order to make intelligence officers more aware of the fact that not every human error is a black swan. Thus, they will be able to identify more accurately a possible 'black swan' and to know how to analyse the concept applied to the intelligence domain.

What is the concept of '*black swan*'? An event can be categorized as a 'black swan' if it meets the following conditions: a) it is an outlier; b) has an extreme impact; c) is explainable and predictable after its occurrence. (Nassim Nicholas Taleb, 2007, p. XVII-XVIII).

According to Cambridge Dictionary '*outlier*' means: 1. '*a person, thing, or fact that is very different from other people, things, or facts, so that it cannot be used to draw general conclusions*'; and 2. '*a fact, figure, piece of data, etc. that is very different from all the others in a set and does not seem to fit the same pattern*' ('Documentation', n.d.).

Therefore, in the analysis of black swan type events, the 'behavioral pattern' of human beings common to past events similar or related to the black swan type that occurred at a given time must be taken into account. For example, if there have been events x at least similar to event X; 'x' means minor damage and 'X' means major damage in all areas or in a certain area. This can be an example of classification. Second, while general conclusions cannot be drawn from the results of a black swan event analysis, some basic guidelines common to likely future black swan events can be established.

In fact, the 'black swan' is an event conceptualized by the author Nassim Nicholas Taleb. On Earth there are events made by human beings and others that are made by God, known as the Great Creator of Universe or the Supreme Force. In my opinion – the 'black swan' – is a cyclical event, that occurs 'naturally' or 'artificially', to realize a change or more, and it is initiated by shadow leaders to establish a new world order whenever it is necessary. Nothing on Earth is left to chance by shadow leaders. Therefore, the architecture of this event takes into account a cyclical time until it occurs, in the case it is created naturally, and if it is artificially created, it must not be used so often, because it will no longer meet its three main characteristics. The shadow leaders rely on these: 'rarity', 'extreme impact' and 'retrospective predictability' (Nassim Nicholas Taleb, 2007, p. XVIII) – to really hide their goals. For the masses, black swans are unexpected, inconceivable events, but for the initiates, they represent normal and necessary events.

Nassim Nicholas Taleb's statement that 'wars are fundamentally unpredictable' (Nassim Nicholas Taleb, 2007, p. XX) is partially wrong. I consider that if Nassim Nicholas Taleb's statement is considered as a truism, it would mean that most of the secret services in this world do not fulfill even the minimum duties, excluding secret missions on foreign territory, neutral or on their own territory. Wars are initiated and conceived with a well-defined purpose, excluding the human being's primitive emotions and ego, as well as the animalic desire to dominate as much as possible, although these have also been a trigger for wars that have marked mankind in the past centuries.

The author uses the metaphor within a story of the same name to describe the abstract and imprecise uncertainty, according to his statement: 'this book is a story, and I prefer to use stories and vignettes to illustrate our gullibility about stories and our preference for the dangerous compression of narratives' (Nassim Nicholas Taleb, 2007, p. XXVII).

In view of the above, a book can be considered a 'black swan' if it can influence and manipulate a mass of people belonging to an age group or more age groups, distracting them from the real problems of society, changing their way of thinking and behavior, feeding their fantasies and desires and channelling their mental energy to their own beliefs. How can this be measured? Through various market studies on the behaviours of human beings.

In order to make a decision you have to take into account the central idea of uncertainty which is to focus on the consequences (that you can know) rather than the probability (which you cannot know) (Nassim Nicholas Taleb, 2007, p. 211).

This is partly true because when a decision is to be made it must be taken into account both the consequences and unforeseen situations that may arise from the decision taken, otherwise the management analysis is incomplete. The unforeseen situations cannot be fully known, but depending on the professional training, the working field, the level at which the decision is taken, the position held, the psycho-social and political order, they can be broadly portrayed in the act of management, in the management plan, in order to achieve maximum results using optimal resources. It is advisable *'to take risks that you can measure than to measure the risks you are taking'* (Nassim Nicholas Taleb, 2010, p. 359).

Good governance means that where appropriate, between two harmful decisions, in the short or medium term, the one that causes the least damage to the public or private entity is chosen, unless there is another solution.

An example that reveals the tendency of every human being to put their own fingerprint is to make a cake: *'people who like to cook, do not make a cake just for the sake of replicating a recipe, they try to make their cake using the ideas of others to improve it'* (Nassim Nicholas Taleb, 2010, p. 230). So, the problem is that the steps of the old recipe are not preserved in making a cake, bringing a note of originality by applying their own ideas to improve it, to adapt it to their appetites.

Similarly, the same problem can occur with information notes or intelligence analyses, generated by those who write them being tempted to put their own mark by distorting the reality and making their own contribution both in the formulation of conclusions and decisions, as well as in the projection of perspectives on a particular area or sphere of activity. The so-called 'improvement' is achieved by its own ideas added 'forced' to the information notes/intelligence analysis. This can happen for fear of failure or for professional reasons, it is known that everyone wants to be in office as much as possible; or for personal reasons to prove that it is still good after a long period of 'professional apathy'. Same event can occur between low–level managers and middle–level managers, between middle–level managers and top–level managers or between low–level managers and top-level managers. Analog this can occur between execution personnel and low–level managers/middle–level managers/top-level managers. Therefore, verification and testing must be done not only horizontally or vertically but also cross-checked within a bureau/department/section/division.

An event is a 'black swan' or 'gray swan' for the great mass of people, but for a small percentage of people it is a normal thing, and for the people who create it, it is just a way to test people or to update the order. What you don't know surprises you.

Like giving out *'antibiotics when it is not very necessary to human beings to make them more vulnerable to severe epidemics'* (Nassim Nicholas Taleb, 2010, p. 328), from a psychological point of view, certain ideas/conceptions/rumours/gossips should be given out to human beings periodically in order to prepare the 'ground' for a particular purpose. In this way, the minds of human beings is prepared so that the desired change does not degenerate into internal uprisings or civil wars.

Nassim Nicholas Taleb argues that his *'results were that regular events can predict regular events, but that extreme events, perhaps because they are more acute when people are unprepared, are almost never predicted from narrow reliance on the past'* (Nassim Nicholas Taleb, 2010, pp. 340-341). I consider that the conclusions presented by the author, Nassim Nicholas Taleb, are partly erroneous. To predict certain events regardless of their impact, a history is needed. Their precise impact cannot be accurately predicted, but based on the history of past events, the nature and source of the impact can be broadly sketched. Thus, in order to create a black swan event, it must be initiated, developed and exploited only when it is absolutely necessary, once time in a few years, to change, for example, a network/intelligence cell in a particular sector of activity, or to achieve a specific purpose, depending on the objective of the mission. A black swan event cannot be created daily or weekly because the frequency would be too high and the black swan would no longer be a black swan event.

It is difficult to predict a particular event and how often it will occur. But, some elements can give you the big picture, and others can narrow your area. Can a human error be a black swan? The nature of the human being is subject to error. In certain sectors of national strategic importance, holding key positions in the institutional, public or private system, human error can be a black swan if it meets the three basic conditions.

The use of black swan in secret services activities

Another objective of this paper is to present the sensitive areas vulnerable to a 'black swan' event that may suddenly occur, and to classify the types of a 'black swan'.

The secret services mainly obtain information through various and complex methods, means and procedures by intelligence officers from operative or by their networks of informants or in case of high ranking intelligence officers from by their agents. These primarily information should be labelled as raw intelligence, but in their usage can occur events such as 'black swans'. That is why in verifying the intelligence analysis the existence or absence of links between what are the objectives of the mission (briefing) and what information is being provided (debriefing) must be taken into account.

What is a 'black swan' in the activities and missions of a secret service? Depending on the type of activities or missions, from case to case, it can be in the operative side and/or on the analysis side (intelligence). Therefore, in the daily papers and the daily activities and missions of the secret services, 'black swan' events can be classified according to where they may occur, as follows:

I. In the Papers:

1) Information notes: a) whose information is taken as 'true' and in the end they prove to be false or non-compliant with reality; b) distortion of information in informa-

tion notes, on request or of their own free will, or misinterpretation of the analyst or intelligence officer in the operative.

2) Human resource assessment:

   a) Internal staff evaluations (certain sensitive elements can be hidden or covered up about certain human beings: physical diseases; mental illness; vices);

   b) Assessments of intelligence officers/agents/informants regarding: b¹) the 'assets' of the activities and missions of the secret services; b²) the individual(s) to be recruited as an informant, agent or intelligence officer. Recruitment can be from internal or external source. Recruitment even if it is from its own schools may have shortcomings in staff assessments.

   c) Assessments of individual 'X' in order to occupy certain functions at national level.

   d) Assessments of individual 'X' in order to occupy certain functions at international level.

II. In the activities and missions of the secret services:

1) an operation or mission can be a 'black swan' if its effects produce huge logistical or informative 'damage' to another secret service or other secret services;

2) the disappearance of high valuable 'assets' of a secret service or several secret services;

3) blocking documents within the secret service or within the circuit of

the National Intelligence Community which would lead to the cessation of certain important operations/missions;

4) hacking the databases of a secret service by another secret service or by several secret services. A hacker or a group of hackers who hack into secret services databases can be a 'black swan'; what is the stake? The dispute between the secret services is not only the identification of foreign agents/intelligence officers/informants, but the confirmation that those identified are what they are to be at the first assessment. That is the one of the reasons the back-up network(s) is (are) very important for the existence of a secret service. The 'hacking' of databases can be a decoy of a secret service for another secret service or for multiple services or for lone hackers;

5) disclosure of the 'safe houses' of a secret service or several secret services;

6) a power blackout or a telecommunications outage. There is a high risk of losing secret documents if the servers are burned (due to sudden major differences in electric current intensity) and when the internet connection is stopped the communication of documents at national or international level is blocked;

7) health crisis – the sudden illness of intelligence officers, agents or informants;

8) a terrorist attack may be a 'black swan';

9) exchange of intelligence. The intelligence are to mislead from the real image. Can be an exchange of intelligence between secret services from the same country or with secret services from other countries.

A 'black swan' can be an event made by an intelligence officer or an informant, a double agent or others human beings used by secret services. Therefore, a 'black swan' can be made intentionally or there may be cases where unintentionally they are made by a human error, a human mislead and so on. To measure a black swan it must be analysed the effects and the changes that it has made. But first step is to identify the source of a 'black swan'. By using network science in identifying the black swan, the search area can be narrowed, thus obtaining some key points in the analysis of its source.

It can be argued that human personnel make the difference between secret services. That is why the process of recruiting intelligence officers/agents requires a lot of attention and competent, objective and visionary people to achieve this sensitive process. Recruiting officers are an important part of the human staff of the secret service.

Legality and morality: In the secret world of secret services, everything is allowed if it is legal or at the limit of legality. Not everything is moral. Morality is a 'coat' that everyone wears, and this raises the question of what is moral and what is not when you have to defend a country and millions of souls?

Post-black swan actions

The post-black-swan reconstruction of a secret service must take primarily the following actions:

a) Minimizing the impact and damage caused by the 'black swan';

b) Identification and analysis of 'black swan';

c) Assessment of staff and existing secret service networks and implement coercive measures.

Other important issues that a secret service may encounter are:

The dissolution and the re-establishment (demilitarization) of a secret service can be a 'black swan' for the great mass of people; the departure of people to the private sector; the transfer or the relocation of human beings from one secret service to another secret service. This raises the question, how do you test the human beings who transfer or relocate from one secret service to another secret service?

Another important and sensitive issue is the existence of 'canned' agents or officers who can work for one, two or more secret services. They can be implanted in national territory or in foreign territory, and used when it is necessary. The challenge is how you identify them.

Conclusion

A first principle to be used in analyzing the origin of a black swan event would be the *cause* and *effect* principle. Each network of a secret service must have a back-up network to withstand a black swan event, which can have its source from inside, from outside, or from other networks. Network science can help to narrow the area of search for the source of a 'black swan', but it will not offer 100% of the right source. Other steps must be taken to increase the percentage provided by network science.

An important process in verifying the accuracy of information obtained by classical methods, means, and procedures would be the periodic evaluation of sources by intelligence officers or counterintelligence officers. Cross-checking can also be the guaran-

tor of information with a high degree of credibility.

The use of a 'black swan' event can be used to verify the trust-authority report and therefore the loyalty of intelligence officers and agents, and so on. Is it good or bad to use such a thing? There are numerous methods, means, and procedures that secret services use to achieve their goals, and it all depends on which side you are on or choose to be on.

Finally, the information presented is not the ultimate solution, but it can offer possible key basic points to prevent the occurrence of a 'black swan' event, or if its occurrence cannot be avoided, it can mitigate its damage.

References

Documentation. (n.d.). In *dictionary.cambridge.org*. Retrieved from https://dictionary.cambridge.org/dictionary/english/outlier

DOD Dictionary of Military and Associated Terms. As of June 2020. Accessed September 5, 2020. https://www.jcs.mil/Portals/36/Documents/Doctrine/pubs/dictionary.pdf.

Hans Bord and Aidan Wilss. 2012. *Overseeing Intelligence Services. A toolkit*, Published by DCAF, Geneva. Accessed August 20, 2020. https://www.dcaf.ch/sites/default/files/publications/documents/Born_Wills_Intelligence_oversight_TK_EN_0.pdf

Intelligence Studies: Types of Intelligence Collection. Accessed September 6, 2020. https://usnwc.libguides.com/c.php?g=494120&p=3381426

Nassim Nicholas Taleb, 2007. *The Black Swan. The Impact of the Highly Improbable*, Random House, New York.

Nassim Nicholas Taleb, 2010. *The Black Swan. The Impact of the Highly Improbable*, Random House Trade Paperback Edition.

Scientific Report no. 2, date 11 January 2019, National Defense University 'Carol I', Bucharest, Romania, 2019.

Romeo-Ionuţ Repez holds a PhD in Intelligence and National Security and an MA in Crisis Management and Conflict Prevention. His research is focused on urban security in the context of identity differences and adds the use of certain elements of network science into a case study to realize a 'coup of city', based on Edward Luttwak's model of a *coup d'état*, which generates the need to create an urban civil defense or urban military defense strategy for specialized public services.

Policy Studies Organization

The Policy Studies Organization (PSO) is a publisher of academic journals and book series, sponsor of conferences, and producer of programs.

Policy Studies Organization publishes dozens of journals on a range of topics, such as European Policy Analysis, Journal of Elder Studies, Indian Politics & Polity, Journal of Critical Infrastructure Policy, and Popular Culture Review.

Additionally, Policy Studies Organization hosts numerous conferences. These conferences include the Middle East Dialogue, Space Education and Strategic Applications Conference, International Criminology Conference, Dupont Summit on Science, Technology and Environmental Policy, World Conference on Fraternalism, Freemasonry and History, and the Internet Policy & Politics Conference.

For more information on these projects, access videos of past events, and upcoming events, please visit us at:

www.ipsonet.org

Related Titles from Westphalia Press

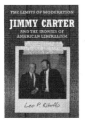

The Limits of Moderation: Jimmy Carter and the Ironies of American Liberalism

The Limits of Moderation: Jimmy Carter and the Ironies of American Liberalism is not a finished product. And yet, even in this unfinished stage, this book is a close and careful history of a short yet transformative period in American political history, when big changes were afoot.

The Zelensky Method
by Grant Farred

Locating Russian's war within a global context, The Zelensky Method is unsparing in its critique of those nations, who have refused to condemn Russia's invasion and are doing everything they can to prevent economic sanctions from being imposed on the Kremlin.

Sinking into the Honey Trap: The Case of the Israeli-Palestinian Conflict
by Daniel Bar-Tal, Barbara Doron, Translator

Sinking into the Honey Trap by Daniel Bar-Tal discusses how politics led Israel to advancing the occupation, and of the deterioration of democracy and morality that accelerates the growth of an authoritarian regime with nationalism and religiosity.

Essay on The Mysteries and the True Object of The Brotherhood of Freemasons
by Jason Williams

The third edition of Essai sur les mystères discusses Freemasonry's role as a society of symbolic philosophers who cultivate their minds, practice virtues, and engage in charity, and underscores the importance of brotherhood, morality, and goodwill.

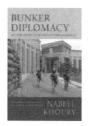

Bunker Diplomacy: An Arab-American in the U.S. Foreign Service
by Nabeel Khoury

After twenty-five years in the Foreign Service, Dr. Nabeel A. Khoury retired from the U.S. Department of State in 2013 with the rank of Minister Counselor. In his last overseas posting, Khoury served as deputy chief of mission at the U.S. embassy in Yemen (2004-2007).

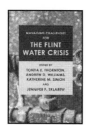

Managing Challenges for the Flint Water Crisis
Edited by Toyna E. Thornton, Andrew D. Williams, Katherine M. Simon, Jennifer F. Sklarew

This edited volume examines several public management and intergovernmental failures, with particular attention on social, political, and financial impacts. Understanding disaster meaning, even causality, is essential to the problem-solving process.

User-Centric Design
by Dr. Diane Stottlemyer

User-centric strategy can improve by using tools to manage performance using specific techniques. User-centric design is based on and centered around the users. They are an essential part of the design process and should have a say in what they want and need from the application based on behavior and performance.

Masonic Myths and Legends
by Pierre Mollier

Freemasonry is one of the few organizations whose teaching method is still based on symbols. It presents these symbols by inserting them into legends that are told to its members in initiation ceremonies. But its history itself has also given rise to a whole mythology.

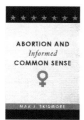

Abortion and Informed Common Sense
by Max J. Skidmore

The controversy over a woman's "right to choose," as opposed to the numerous "rights" that abortion opponents decide should be assumed to exist for "unborn children," has always struck me as incomplete. Two missing elements of the argument seems obvious, yet they remain almost completely overlooked.

The Athenian Year Primer: Attic Time-Reckoning and the Julian Calendar
by Christopher Planeaux

The ability to translate ancient Athenian calendar references into precise Julian-Gregorian dates will not only assist Ancient Historians and Classicists to date numerous historical events with much greater accuracy but also aid epigraphists in the restorations of numerous Attic inscriptions.

Siddhartha: Life of the Buddha
by David L. Phillips,
contributions by Venerable Sitagu Sayadaw

Siddhartha: Life of the Buddha is an illustrated story for adults and children about the Buddha's birth, enlightenment and work for social justice. It includes illustrations from Pagan, Burma which are provided by Rev. Sitagu Sayadaw.

Growing Inequality: Bridging Complex Systems, Population Health, and Health Disparities
Editors: George A. Kaplan, Ana V. Diez Roux, Carl P. Simon, and Sandro Galea

Why is America's health is poorer than the health of other wealthy countries and why health inequities persist despite our efforts? In this book, researchers report on groundbreaking insights to simulate how these determinants come together to produce levels of population health and disparities and test new solutions.

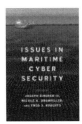

Issues in Maritime Cyber Security
Edited by Dr. Joe DiRenzo III, Dr. Nicole K. Drumhiller, and Dr. Fred S. Roberts

The complexity of making MTS safe from cyber attack is daunting and the need for all stakeholders in both government (at all levels) and private industry to be involved in cyber security is more significant than ever as the use of the MTS continues to grow.

Female Emancipation and Masonic Membership:
An Essential Collection
By Guillermo De Los Reyes Heredia

Female Emancipation and Masonic Membership: An Essential Combination is a collection of essays on Freemasonry and gender that promotes a transatlantic discussion of the study of the history of women and Freemasonry and their contribution in different countries.

Anti-Poverty Measures in America: Scientism and Other Obstacles
Editors, Max J. Skidmore and Biko Koenig

Anti-Poverty Measures in America brings together a remarkable collection of essays dealing with the inhibiting effects of scientism, an over-dependence on scientific methodology that is prevalent in the social sciences, and other obstacles to anti-poverty legislation.

Geopolitics of Outer Space: Global Security and Development
by Ilayda Aydin

A desire for increased security and rapid development is driving nation-states to engage in an intensifying competition for the unique assets of space. This book analyses the Chinese-American space discourse from the lenses of international relations theory, history and political psychology to explore these questions.

Contests of Initiative: Countering China's Gray Zone Strategy in the East and South China Seas
by Dr. Raymond Kuo

China is engaged in a widespread assertion of sovereignty in the South and East China Seas. It employs a "gray zone" strategy: using coercive but sub-conventional military power to drive off challengers and prevent escalation, while simultaneously seizing territory and asserting maritime control.

Discourse of the Inquisitive
Editors: Jaclyn Maria Fowler and Bjorn Mercer

Good communication skills are necessary for articulating learning, especially in online classrooms. It is often through writing that learners demonstrate their ability to analyze and synthesize the new concepts presented in the classroom.

westphaliapress.org

This publication is available open access at:

https://gsis.scholasticahq.com/

http://www.ipsonet.org/publications/open-access

Thanks to the generosity of the American Public University System